Essential Cell Biology

Essential Cell Biology

Philip Newsom

R CALLISTO REFERENCE

www.callistoreference.com

Callisto Reference,
118-35 Queens Blvd., Suite 400,
Forest Hills, NY 11375, USA

Visit us on the World Wide Web at:
www.callistoreference.com

ISBN: 978-1-64116-557-0 (Hardback)

Cataloging-in-Publication Data

Essential cell biology / Philip Newsom.
     p. cm.
Includes bibliographical references and index.
ISBN 978-1-64116-557-0
1. Cytology. 2. Cells. 3. Biology. I. Newsom, Philip.
QH581.2 .E87 2022
571.6--dc23

Table of Contents

Permissions

Index

Preface

The branch of biology which deals with the study of the function and structure of the cell is referred to as cell biology. It focuses on the physiological attributes, metabolic processes and life cycle of the cell. It also studies the signaling pathways, chemical composition and interactions of the cell with their environment. It encompasses both prokaryotic and eukaryotic cells. Research in this field is conducted on both microscopic and molecular levels. The knowledge of the components of cells as well as their functioning plays a vital role in the studies related to different bio-medical fields. This textbook is a valuable compilation of topics, ranging from the basic to the most complex theories and principles in the field of cell biology. Some of the diverse topics covered herein address the varied branches that fall under this category. This book will provide comprehensive knowledge to the readers.

A short introduction to every chapter is written below to provide an overview of the content of the book:

Chapter 1 - Cell biology is a domain of biology that is concerned with the study of the structure and function of the cell. It primarily deals with the metabolic processes, lifecycle, physiological properties and chemical composition of the cell. This chapter has been carefully written to provide an easy introduction to these facets of cell biology.; **Chapter 2** - Cells are composed of various macromolecules such as nucleic acid, amino acids and proteins. Besides these, the extracellular macromolecules which are present outside the cells provide biochemical and structural support to the cells around them. The network of these molecules is known as the extracellular matrix. This chapter has been carefully written to provide an easy understanding of these macromolecules in cells.; **Chapter 3** - The anatomy of the cell is made up of a number of components such as cell membrane, cell organelles, cell wall, cytoplasm, cytoskeleton and cytosol. The topics elaborated in this chapter will help in gaining a better perspective about the anatomy of the cell as well as these components.; **Chapter 4** - The communication process which governs basic activities of cells and coordinates multiple cell actions is known as cell signaling. They are classified as either biochemical or mechanical in nature, based on the kind of signal. This chapter discusses in detail the processes and concepts related to cell signaling and communication.; **Chapter 5** - There are several biological activities which take place within cells, such as respiration, reproduction and migration. The metabolic reactions and processes through which cells transform biochemical energy from nutrients into adenosine triphosphate are known as cell respiration. All these diverse biological activities of the cells have been carefully analyzed in this chapter.

Finally, I would like to thank my fellow scholars who gave constructive feedback and my family members who supported me at every step.

Philip Newsom

Chapter 1

An Introduction to Cells and Cell Biology

Cell biology is a domain of biology that is concerned with the study of the structure and function of the cell. It primarily deals with the metabolic processes, lifecycle, physiological properties and chemical composition of the cell. This chapter has been carefully written to provide an easy introduction to these facets of cell biology.

Cells are the basic unit of life. In the modern world, they are the smallest known world that performs all of life's functions. All living organisms are either single cells, or are multicellular organisms composed of many cells working together.

Cells are the smallest known unit that can accomplish all of these functions. Defining characteristics that allow a cell to perform these functions include:

- A cell membrane that keeps the chemical reactions of life together.

- At least one chromosome, composed of genetic material that contain the cell's "blueprints" and "software."

- Cytoplasm – the fluid inside the cell, in which the chemical processes of life occur.

Function of Cells

Scientists define seven functions that must be fulfilled by a living organism. These are:

- A living thing must respond to changes in its environment.

- A living thing must grow and develop across its lifespan.

- A living thing must be able to reproduce, or make copies of itself.

- A living thing must have metabolism.

- A living thing must maintain homeostasis, or keep its internal environment the same regardless of outside changes.

- A living thing must be made of cells.

- A living thing must pass on traits to its offspring.

It is the biology of cells which enables living things to perform all of these functions. Below, we discuss how they make the functions of life possible.

Working of Cells

In order to accomplish them, they must have:

- A cell membrane that separates the inside of the cell from the outside. By concentrating the chemical reactions of life inside a small area within a membrane, cells allow the reactions of life to proceed much faster than they otherwise would.

- Genetic material which is capable of passing on traits to the cell's offspring. In order to reproduce, organisms must ensure that their offspring have all the information that they need to be able to carry out all the functions of life. All modern cells accomplish this using DNA, whose base-pairing properties allow cells to make accurate copies of a cell's "blueprints" and "operating system." Some scientists think that the first cells might have used RNA instead.

- Proteins that perform a wide variety of structural, metabolic, and reproductive functions. There are countless different functions that cells must perform to obtain energy and reproduce.

Depending on the cell, examples of these functions can include photosynthesis, breaking down sugar, locomotion, copying its own DNA, allowing certain substances to pass through the cell membrane while keeping others out, etc.

Proteins are made of amino acids, which are like the "Legos" of biochemistry. Amino acids come in different sizes, different shapes, and with different properties such as polarity, ionic charge, and hydrophobicity.

By putting amino acids together based on the instructions in their genetic material, cells can create biochemical machinery to perform almost any function.

Some scientists think that the first cells might have used RNA to accomplish some vital functions, and then moved to much more versatile amino acids to do the job as the result of a mutation.

The different cell types we will discuss have different ways of accomplishing these functions:

Prokaryotes

Prokaryotes are the simpler and older of the two major types of cells. Prokaryotes are single-celled organisms. Bacteria and archaebacteria are examples of prokaryotic cells.

Prokaryotic cells have a cell membrane, and one or more layers of additional protection from the outside environment. Many prokaryotes have a cell membrane made of phospholipids, enclosed by a cell wall made of a rigid sugar. The cell wall may be enclosed by another thick "capsule" made of sugars.

Many prokaryotic cells also have cilia, tails, or other ways in which the cell can control its movement.

These characteristics, as well as the cell wall and capsule, reflect the fact that prokaryotic cells are going it alone in the environment. They are not part of a multicellular organism, which might have whole layers of cells devoted to protecting other cells from the environment, or to creating motion.

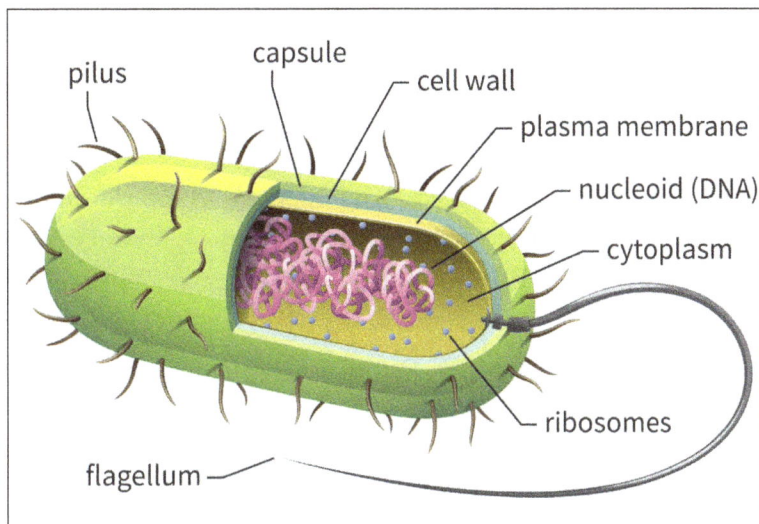

Prokaryote cell

Prokaryotic cells have a single chromosome which contains all of the cell's essential hereditary material and operating instructions. This single chromosome is usually round. There is no nucleus, or any other internal membranes or organelles. The chromosome just floats in the cell's cytoplasm.

Additional genetic traits and information might be contained in other gene units within the cytoplasm, called "plasmids," but these are usually genes that are passed back and forth by prokaryotes though the process of "horizontal gene transfer," which is when one cell gives genetic material to another. Plasmids contain non-essential DNA that the cell can live without, and which is not necessarily passed on to offspring.

When a prokaryotic cell is ready to reproduce, it makes a copy of its single chromosome. Then the cell splits in half, apportioning one copy of its chromosome and a random assortment of plasmids to each daughter cell.

There are two major types of prokaryotes known to scientists to date: archaebacteria, which are a very old lineage of life with some biochemical differences from bacteria and eukaryotes, and bacteria, sometimes called "eubacteria," or "true bacteria" to differentiate them from archaebacteria.

Bacteria are thought to be more "modern" descendants of archaebacteria. Both families have "bacteria" in the name because the differences between them were not understood prior to the invention of modern biochemical and genetic analysis techniques.

When scientists began to examine the biochemistry and genetics of prokaryotes in detail, they discovered these two very different groups, who probably have different relationships to eukaryotes and different evolutionary histories. Some scientists think that eukaryotes like humans are more closely related to bacteria, since eukaryotes have similar cell membrane chemistry to bacteria. Others think that archaebacteria are more closely related to us eukaryotes, since they use similar proteins to reproduce their chromosomes.

Still others think that we might be descended from both – that eukaryotic cells might have come into existence when archaebacteria started living inside of a bacterial cell, or vice versa. This would explain how we have important genetic and chemical attributes of both, and why we have multiple internal compartments such as the nucleus, chloroplasts, and mitochondria.

Eukaryotes

Eukaryotic cells are thought to be the most modern major cell type. All multicellular organisms, including you, your cat, and your houseplants, are eukaryotes. Eukaryotic cells seem to have "learned" to work together to create multicellular organisms, while prokaryotes seem unable to do this.

Eukaryotic cells usually have more than one chromosome, which contains large amounts of genetic information. Within the body of a multicellular organism, different genes within these chromosomes may be switched "on" and "off," allowing for cells that have different traits and perform different functions within the same organism.

Eukaryotic cells also have one or more internal membranes, which has led scientists to the conclusion that eukaryotic cells likely evolved when one or more types of prokaryote began living in symbiotic relationships inside of other cells.

Organelles with interior membranes found in eukaryotic cells typically include:

- For animal cells: Mitochondria, which liberate the energy from sugar and turn it into ATP in an extremely efficient way. Mitochondria even have their own DNA, separate from the cells' nuclear DNA, which gives further support for the theory that they used to be independent bacteria.

- For plant cells: Chloroplasts, which perform photosynthesis, making ATP and sugar from sunlight and air. Chloroplasts also have their own DNA, suggesting that they may have originated as photosynthetic bacteria.

- Nucleus: In eukaryotic cells, the nucleus contains the essential DNA blueprints and operating instructions for the cell. The nuclear envelope is thought to provide an extra layer of protection for the DNA against toxins or invaders which might damage it. It is unknown whether the nucleus might also have been an endosymbiotic prokaryote at one time, or whether its membrane simply evolved as an extra layer of protection for the cell's DNA.

- Endoplasmic reticulum: This complex internal membrane is a major site of protein creation for cells. The evolutionary origin of the endoplasmic reticulum is not known.

- Golgi apparatus: This internal membrane complex can be thought of like the endoplasmic reticulum's "post office." It receives proteins from the ER, packages and "labels" them by attaching sugars as needed, and then ships them off to their final destinations.

- Others: Many eukaryotic cells can create temporary internal membrane "sacs," called "vacuoles," to store waste, or to package important materials.

Some cells, for example have special vacuoles called "lysosomes" which are full of corrosive substances and digestive enzymes. Cells simply dump their "trash" into lysosomes, where the harsh environment breaks them down into simpler components that can be re-used.

Examples of Cells

Archaebacteria

Archaebacteria are a very old form of prokaryotic cells. Biologists actually put them in their own "domain" of life, separate from other bacteria.

Key ways in which archaebacteria differ from other bacteria includes:

- Their cell membranes, which are made of a type of lipid not found in either bacteria or eukaryotic cell membranes.

- Their DNA replication enzymes, which are more similar to those of eukaryotes than those of bacteria, suggesting that bacteria and archae are only distantly related, and archaebacteria may actually be more closely related to us than to modern bacteria.

- Some archaebacteria have the ability to produce methane, which is a metabolic process not found in any bacteria or any eukaryotes.

Archaebacteria's unique chemical attributes allow them to live in extreme environments, such as superheated water, extremely salty water, and some environments which are toxic to all other life forms.

Scientists became very excited in recent years at the discovery of *Lokiarchaeota* – a type of archaebacteria which shares many genes with eukaryotes that had never before been found in prokaryotic cells. It is now thought that *Lokiarchaeota* may be our closest living relative in the prokaryotic world.

Bacteria

We are most likely familiar with the type of bacteria that can make you sick. Indeed, common pathogens like *Streptococcus*and *Staphylococcus* are prokaryotic bacterial cells.

But there are also many types of helpful bacteria – including those that break down dead waste to turn useless materials into fertile soil, and bacteria that live in our own digestive tract and help us digest food.

Bacterial cells can commonly be found living in symbiotic relationships with multicellular organisms like us, in the soil, and anywhere else that's not too extreme for them to live.

Plant Cells

Plant cells are eukaryotic cells that are part of multicellular, photosynthetic organisms. Plants cells have chloroplast organelles, which contain pigments that absorb photons of light and harvest the energy of those photons.

Chloroplasts have the remarkable ability to turn light energy into cellular fuel, and use this energy to take carbon dioxide from the air and turn it into sugars that can be used by living things as fuel or building material.

In addition to having chloroplasts, plant cells also typically have a cell wall made of rigid sugars, to enable plant tissues to maintain their upright structures such as leaves, stems, and tree trunks.

Plant cells also have the usual eukaryotic organelles including a nucleus, endoplasmic reticulum, and Golgi apparatus.

Animal Cells

Like all animal cells, it has mitochondria which perform cellular respiration, turning oxygen and sugar into large amounts of ATP to power cellular functions.

It also has the same organelles as most animal cells: a nucleus, endoplasmic reticulum, Golgi apparatus, etc. But as part of a multicellular organism, your liver cell also expresses unique genes, which give it unique traits and abilities.

Liver cells in particular contain enzymes that break down many toxins, which is what allows the liver to purify your blood and break down dangerous bodily waste. The liver cell is an excellent example of how multicellular organisms can be more efficient by having different cell types work together.

Our body could not survive without liver cells to break down certain toxins and waste products, but the liver cell itself could not survive without nerve and muscle cells that help you find food, and a digestive tract to break down that food into easily digestible sugars. And all of these cell types contain the information to make all the other cell types! It's simply a matter of which genes are switched "on" or "off" during development.

Eukaryotic Cells

A eukaryotic cell has a true membrane-bound nucleus and has other membranous organelles that allow for compartmentalization of functions.

Eukaryotic Cell Structure

Like a prokaryotic cell, a eukaryotic cell has a plasma membrane, cytoplasm, and ribosomes. However, unlike prokaryotic cells, eukaryotic cells have:

1. A membrane-bound nucleus.

2. Numerous membrane-bound organelles (including the endoplasmic reticulum, Golgi apparatus, chloroplasts, and mitochondria).

3. Several rod-shaped chromosomes.

Because a eukaryotic cell's nucleus is surrounded by a membrane, it is often said to have a "true nucleus." Organelles (meaning "little organ") have specialized cellular roles, just as the organs of your body have specialized roles. They allow different functions to be compartmentalized in different areas of the cell.

The Nucleus and its Structures

Typically, the nucleus is the most prominent organelle in a cell. Eukaryotic cells have a true nucleus, which means the cell's DNA is surrounded by a membrane. Therefore, the nucleus houses the cell's DNA and directs the synthesis of proteins and ribosomes, the cellular organelles responsible for protein synthesis. The nuclear envelope is a double-membrane structure that constitutes the outermost portion of the nucleus. Both the inner and outer membranes of the nuclear envelope are phospholipid bilayers. The nuclear envelope is punctuated with pores that control the passage of ions, molecules, and RNA between the nucleoplasm and cytoplasm. The nucleoplasm is the semi-solid fluid inside the nucleus where we find the chromatin and the nucleolus. Furthermore, chromosomes are structures within the nucleus that are made up of DNA, the genetic material. In prokaryotes, DNA is organized into a single circular chromosome. In eukaryotes, chromosomes are linear structures.

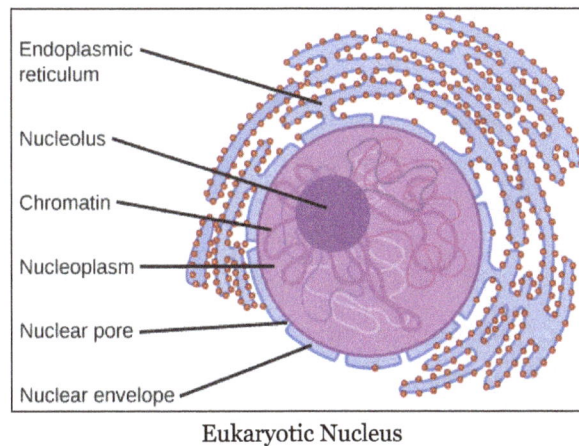

Eukaryotic Nucleus

In figure, the nucleus stores chromatin (DNA plus proteins) in a gel-like substance called the nucleoplasm. The nucleolus is a condensed region of chromatin where ribosome synthesis occurs. The boundary of the nucleus is called the nuclear envelope. It consists of two phospholipid bilayers: an outer membrane and an inner membrane. The nuclear membrane is continuous with the endoplasmic reticulum. Nuclear pores allow substances to enter and exit the nucleus.

Other Membrane-bound Organelles

Mitochondria are oval-shaped, double membrane organelles that have their own ribosomes and DNA. These organelles are often called the "energy factories" of a cell because they are responsible for making adenosine triphosphate (ATP), the cell's main energy-carrying molecule, by conducting cellular respiration. The endoplasmic reticulum modifies proteins and synthesizes lipids, while the golgi apparatus is where the sorting, tagging, packaging, and distribution of lipids and proteins takes place. Peroxisomes are small, round organelles enclosed by single membranes; they carry out oxidation reactions that break down fatty acids and amino acids. Peroxisomes also detoxify many poisons that may enter the body. Vesicles and vacuoles are membrane-bound sacs that function in storage and transport. Other than the fact that vacuoles are somewhat larger than vesicles, there is a very subtle distinction between them: the membranes of vesicles can fuse with either the plasma membrane or other membrane systems within the cell. All of these organelles are found in each and every eukaryotic cell.

Animal Cells versus Plant Cells

While all eukaryotic cells contain the aforementioned organelles and structures, there are some striking differences between animal and plant cells. Animal cells have a centrosome and lysosomes, whereas plant cells do not. The centrosome is a microtubule-organizing center found near the nuclei of animal cells while lysosomes take care of the cell's digestive process.

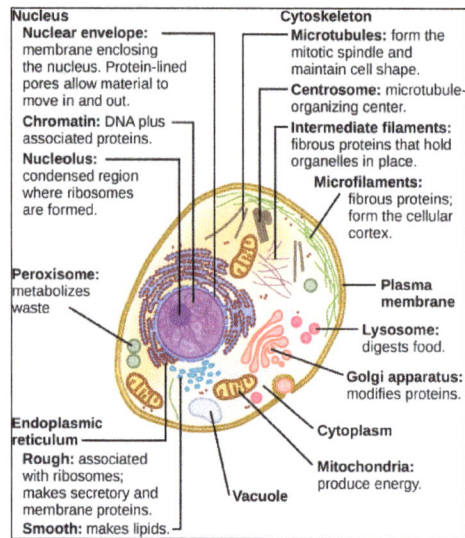

- Animal Cells: Despite their fundamental similarities, there are some striking differences between animal and plant cells. Animal cells have centrioles, centrosomes, and lysosomes, whereas plant cells do not.

In addition, plant cells have a cell wall, a large central vacuole, chloroplasts, and other specialized plastids, whereas animal cells do not. The cell wall protects the cell, provides structural support, and gives shape to the cell while the central vacuole plays a key role in regulating the cell's concentration of water in changing environmental conditions. Chloroplasts are the organelles that carry out photosynthesis.

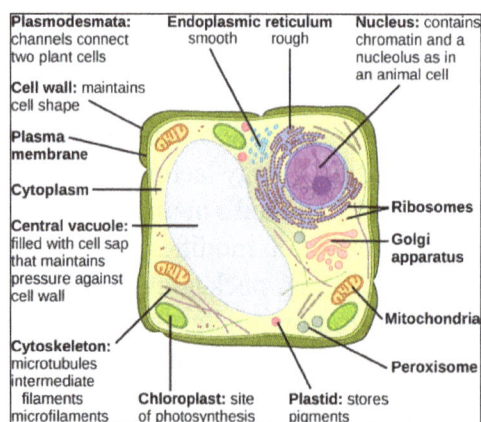

- Plant Cells: Plant cells have a cell wall, chloroplasts, plasmodesmata, and plastids used for storage, and a large central vacuole, whereas animal cells do not.

Plasma Membrane and the Cytoplasm

The plasma membrane is made up of a phospholipid bilayer that regulates the concentration of substances that can permeate a cell.

Plasma Membrane

Despite differences in structure and function, all living cells in multicellular organisms have a surrounding plasma membrane (also known as the cell membrane). As the outer layer of your skin separates your body from its environment, the plasma membrane separates the inner contents of a cell from its exterior environment. The plasma membrane can be described as a phospholipid bilayer with embedded proteins that controls the passage of organic molecules, ions, water, and oxygen into and out of the cell. Wastes (such as carbon dioxide and ammonia) also leave the cell by passing through the membrane.

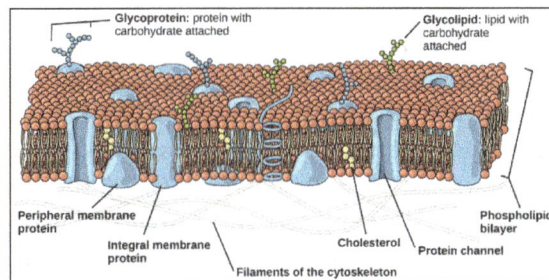

Eukaryotic Plasma Membrane: The eukaryotic plasma membrane is a phospholipid bilayer with proteins and cholesterol embedded in it.

The cell membrane is an extremely pliable structure composed primarily of two adjacent sheets of phospholipids. Cholesterol, also present, contributes to the fluidity of the membrane. A single phospholipid molecule consists of a polar phosphate "head," which is hydrophilic, and a non-polar lipid "tail," which is hydrophobic. Unsaturated fatty acids result in kinks in the hydrophobic tails. The phospholipid bilayer consists of two phospholipids arranged tail to tail. The hydrophobic tails associate with one another, forming the interior of the membrane. The polar heads contact the fluid inside and outside of the cell.

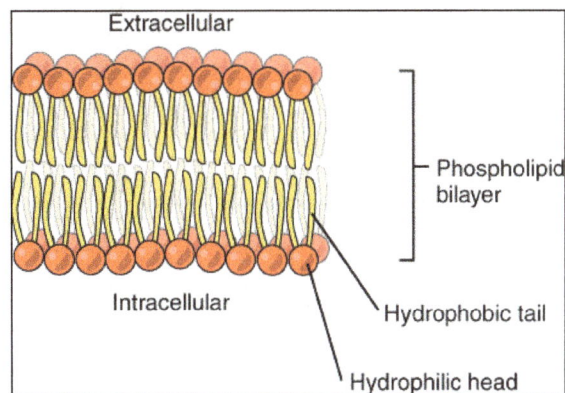

- Phospholipid Bilayer: The phospholipid bilayer consists of two adjacent sheets of phospholipids, arranged tail to tail. The hydrophobic tails associate with one another, forming the interior of the membrane. The polar heads contact the fluid inside and outside of the cell.

The plasma membrane's main function is to regulate the concentration of substances inside the cell. These substances include ions such as Ca^{++}, Na^+, K^+, and Cl^-; nutrients including sugars, fatty acids, and amino acids; and waste products, particularly carbon dioxide (CO_2), which must leave the cell.

The membrane's lipid bilayer structure provides the cell with access control through permeability. The phospholipids are tightly packed together, while the membrane has a hydrophobic interior. This structure causes the membrane to be selectively permeable. A membrane that has selective permeability allows only substances meeting certain criteria to pass through it unaided. In the case of the plasma membrane, only relatively small, non-polar materials can move through the lipid bilayer (remember, the lipid tails of the membrane are nonpolar). Some examples of these materials are other lipids, oxygen and carbon dioxide gases, and alcohol. However, water-soluble materials—such as glucose, amino acids, and electrolytes—need some assistance to cross the membrane because they are repelled by the hydrophobic tails of the phospholipid bilayer.

Transport across the Membrane

All substances that move through the membrane do so by one of two general methods, which are categorized based on whether or not energy is required. Passive (non-energy requiring) transport is the movement of substances across the membrane without the expenditure of cellular energy. During this type of transport, materials move by simple diffusion or by facilitated diffusion through the membrane, down their concentration gradient. Water passes through the membrane in a diffusion process called osmosis. Osmosis is the diffusion of water through a semi-permeable membrane down its concentration gradient. It occurs when there is an imbalance of solutes outside of a cell versus inside the cell. The solution that has the higher concentration of solutes is said to be hypertonic and the solution that has the lower concentration of solutes is said to be hypotonic. Water molecules will diffuse out of the hypotonic solution and into the hypertonic solution (unless acted upon by hydrostatic forces).

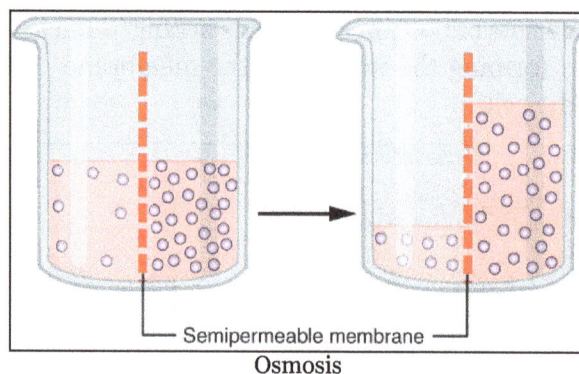

Semipermeable membrane
Osmosis

In figure, psmosis is the diffusion of water through a semipermeable membrane down its concentration gradient. If a membrane is permeable to water, though not to a solute, water will equalize its own concentration by diffusing to the side of lower water concentration (and thus the side of higher solute concentration). In the beaker on the left, the solution on the right side of the membrane is hypertonic.

In contrast to passive transport, active (energy-requiring) transport is the movement of substances across the membrane using energy from adenosine triphosphate (ATP). The energy is expended to assist material movement across the membrane in a direction against their concentration gradient. Active transport may take place with the help of protein pumps or through the use of vesicles. Another form of this type of transport is endocytosis, where a cell envelopes extracellular materials using its cell membrane. The opposite process is known as exocytosis. This is where a cell exports material using vesicular transport.

Cytoplasm

The cell's plasma membrane also helps contain the cell's cytoplasm, which provides a gel-like environment for the cell's organelles. The cytoplasm is the location for most cellular processes, including metabolism, protein folding, and internal transportation.

Nucleus and Ribosomes

Found within eukaryotic cells, the nucleus contains the genetic material that determines the entire structure and function of that cell.

Nucleus

One of the main differences between prokaryotic and eukaryotic cells is the nucleus. As previously discussed, prokaryotic cells lack an organized nucleus while eukaryotic cells contain membrane-bound nuclei (and organelles) that house the cell's DNA and direct the synthesis of ribosomes and proteins.

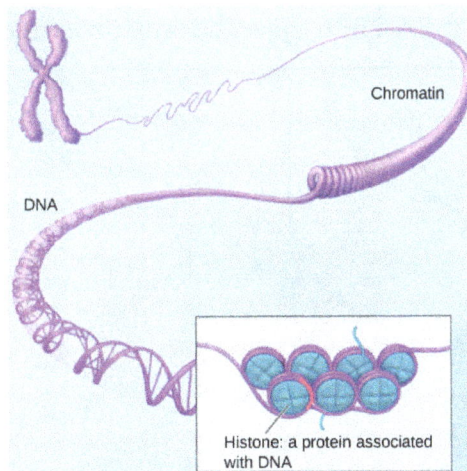

DNA is highly organized: This image shows various levels of the organization of chromatin
(DNA and protein). Along the chromatin threads, unwound protein-chromosome complexes,
we find DNA wrapped around a set of histone proteins.

The nucleus stores chromatin (DNA plus proteins) in a gel-like substance called the nucleoplasm. To understand chromatin, it is helpful to first consider chromosomes. Chromatin describes the material that makes up chromosomes, which are structures within the nucleus that are made up of DNA, the hereditary material. You may remember that in prokaryotes, DNA is organized into a single circular chromosome. In eukaryotes, chromosomes are linear structures. Every eukaryotic species has a specific number of chromosomes in the nuclei of its body's cells. For example, in humans, the chromosome number is 46, while in fruit flies, it is eight. Chromosomes are only visible and distinguishable from one another when the cell is getting ready to divide. In order to organize the large amount of DNA within the nucleus, proteins called histones are attached to chromosomes; the DNA is wrapped around these histones to form a structure resembling beads on a string. These protein-chromosome complexes are called chromatin.

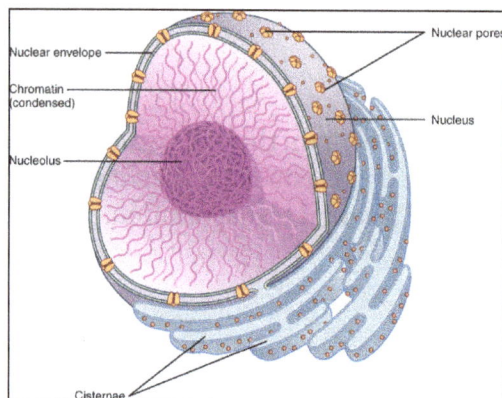

The nucleus stores the hereditary material of the cell: The nucleus is the control center of the cell. The nucleus of living cells contains the genetic material that determines the entire structure and function of that cell.

The nucleoplasm is also where we find the nucleolus. The nucleolus is a condensed region of chromatin where ribosome synthesis occurs. Ribosomes, large complexes of protein and ribonucleic acid (RNA), are the cellular organelles responsible for protein synthesis. They receive their "orders" for protein synthesis from the nucleus where the DNA is transcribed into messenger RNA (mRNA). This mRNA travels to the ribosomes, which translate the code provided by the sequence of the nitrogenous bases in the mRNA into a specific order of amino acids in a protein.

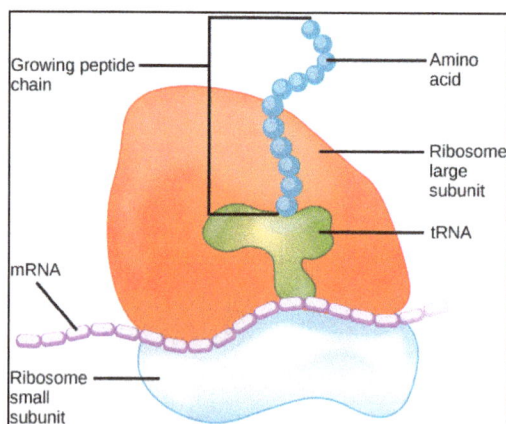

Ribosomes are responsible for protein synthesis: Ribosomes are made up of a large subunit (top) and a small subunit (bottom). During protein synthesis, ribosomes assemble amino acids into proteins.

Lastly, the boundary of the nucleus is called the nuclear envelope. It consists of two phospholipid bilayers: an outer membrane and an inner membrane. The nuclear membrane is continuous with the endoplasmic reticulum, while nuclear pores allow substances to enter and exit the nucleus.

Mitochondria

Mitochondria are organelles that are responsible for making adenosine triphosphate (ATP), the cell's main energy-carrying molecule.

One of the major features distinguishing prokaryotes from eukaryotes is the presence of mitochondria. Mitochondria are double-membraned organelles that contain their own ribosomes and DNA. Each membrane is a phospholipid bilayer embedded with proteins. Eukaryotic cells may contain anywhere from one to several thousand mitochondria, depending on the cell's level of energy consumption. Each mitochondrion measures 1 to 10 micrometers (or greater) in length and exists in the cell as an organelle that can be ovoid to worm-shaped to intricately branched.

Mitochondria Structure

Most mitochondria are surrounded by two membranes, which would result when one membrane-bound organism was engulfed into a vacuole by another membrane-bound organism. The mitochondrial inner membrane is extensive and involves substantial infoldings called cristae that resemble the textured, outer surface of alpha-proteobacteria. The matrix and inner membrane are rich with the enzymes necessary for aerobic respiration.

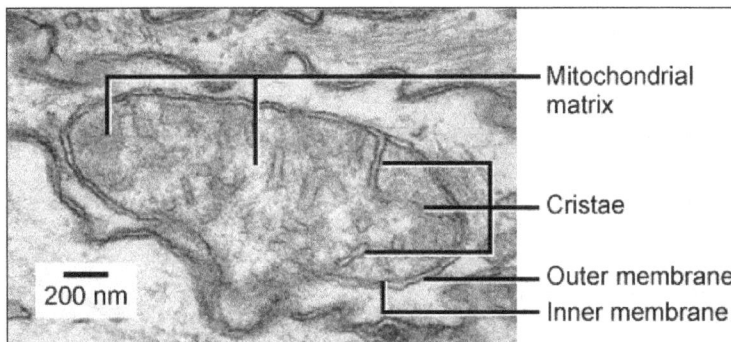

Mitochondrial structure

In figure, this electron micrograph shows a mitochondrion as viewed with a transmission electron microscope. This organelle has an outer membrane and an inner membrane. The inner membrane contains folds, called cristae, which increase its surface area. The space between the two membranes is called the inter-membrane space, and the space inside the inner membrane is called the mitochondrial matrix. ATP synthesis takes place on the inner membrane.

Mitochondria have their own (usually) circular DNA chromosome that is stabilized by attachments to the inner membrane and carries genes similar to genes expressed by alpha-proteobacteria. Mitochondria also have special ribosomes and transfer RNAs that resemble these components in prokaryotes. These features all support the hypothesis that mitochondria were once free-living prokaryotes.

Mitochondria Function

Mitochondria are often called the "powerhouses" or "energy factories" of a cell because they are responsible for making adenosine triphosphate (ATP), the cell's main energy-carrying molecule. ATP represents the short-term stored energy of the cell. Cellular respiration is the process of making ATP using the chemical energy found in glucose and other nutrients. In mitochondria, this process uses oxygen and produces carbon dioxide as a waste product. In fact, the carbon dioxide that you exhale with every breath comes from the cellular reactions that produce carbon dioxide as a by-product.

It is important to point out that muscle cells have a very high concentration of mitochondria that produce ATP. Your muscle cells need a lot of energy to keep your body moving. When your cells don't get enough oxygen, they do not make a lot of ATP. Instead, the small amount of ATP they make in the absence of oxygen is accompanied by the production of lactic acid.

In addition to the aerobic generation of ATP, mitochondria have several other metabolic functions. One of these functions is to generate clusters of iron and sulfur that are important cofactors of many enzymes. Such functions are often associated with the reduced mitochondrion-derived organelles of anaerobic eukaryotes.

Origins of Mitochondria

There are two hypotheses about the origin of mitochondria: endosymbiotic and autogenous, but the most accredited theory at present is endosymbiosis. The endosymbiotic hypothesis suggests mitochondria were originally prokaryotic cells, capable of implementing oxidative mechanisms. These prokaryotic cells may have been engulfed by a eukaryote and became endosymbionts living inside the eukaryote.

Comparing Plant and Animal Cells

Although they are both eukaryotic cells, there are unique structural differences between animal and plant cells.

Animal Cells versus Plant Cells

Each eukaryotic cell has a plasma membrane, cytoplasm, a nucleus, ribosomes, mitochondria, peroxisomes, and in some, vacuoles; however, there are some striking differences between animal and plant cells. While both animal and plant cells have microtubule organizing centers (MTOCs), animal cells also have centrioles associated with the MTOC: a complex called the centrosome. Animal cells each have a centrosome and lysosomes, whereas plant cells do not. Plant cells have a cell wall, chloroplasts and other specialized plastids, and a large central vacuole, whereas animal cells do not.

Centrosome

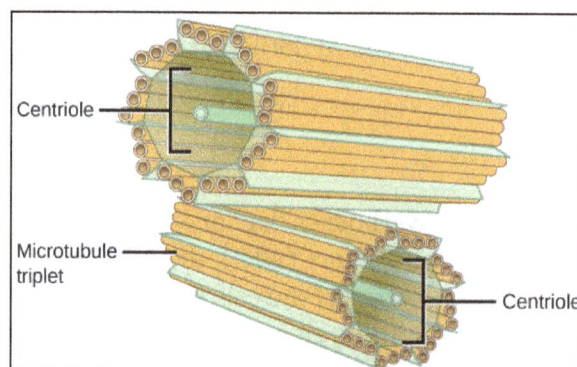

The Centrosome Structure: The centrosome consists of two centrioles that lie at right angles to each other. Each centriole is a cylinder made up of nine triplets of microtubules. Non-tubulin proteins (indicated by the green lines) hold the microtubule triplets together.

The centrosome is a microtubule-organizing center found near the nuclei of animal cells. It contains a pair of centrioles, two structures that lie perpendicular to each other. Each centriole is a cylinder of nine triplets of microtubules. The centrosome (the organelle where all microtubules originate) replicates itself before a cell divides, and the centrioles appear to have some role in pulling the duplicated chromosomes to opposite ends of the dividing cell. However, the exact function of the centrioles in cell division isn't clear, because cells that have had the centrosome removed can still divide; and plant cells, which lack centrosomes, are capable of cell division.

Lysosomes

Animal cells have another set of organelles not found in plant cells: lysosomes. The lysosomes are the cell's "garbage disposal." In plant cells, the digestive processes take place in vacuoles. Enzymes within the lysosomes aid the breakdown of proteins, polysaccharides, lipids, nucleic acids, and even worn-out organelles. These enzymes are active at a much lower pH than that of the cytoplasm. Therefore, the pH within lysosomes is more acidic than the pH of the cytoplasm. Many reactions that take place in the cytoplasm could not occur at a low pH, so the advantage of compartmentalizing the eukaryotic cell into organelles is apparent.

Cell Wall

The cell wall is a rigid covering that protects the cell, provides structural support, and gives shape to the cell. Fungal and protistan cells also have cell walls. While the chief component of prokaryotic cell walls is peptidoglycan, the major organic molecule in the plant cell wall is cellulose, a polysaccharide comprised of glucose units. When you bite into a raw vegetable, like celery, it crunches. That's because you are tearing the rigid cell walls of the celery cells with your teeth.

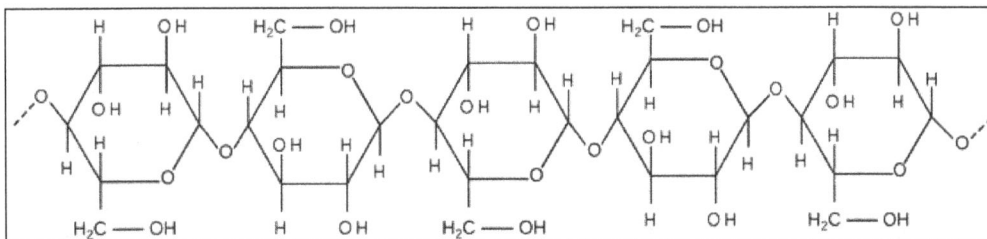

Cellulose: Cellulose is a long chain of β-glucose molecules connected by a 1-4 linkage.
The dashed lines at each end of the figure indicate a series of many more glucose units.
The size of the page makes it impossible to portray an entire cellulose molecule.

Chloroplasts

Like mitochondria, chloroplasts have their own DNA and ribosomes, but chloroplasts have an entirely different function. Chloroplasts are plant cell organelles that carry out photosynthesis. Photosynthesis is the series of reactions that use carbon dioxide, water, and light energy to make glucose and oxygen. This is a major difference between plants and animals; plants (autotrophs) are able to make their own food, like sugars, while animals (heterotrophs) must ingest their food.

Like mitochondria, chloroplasts have outer and inner membranes, but within the space enclosed by a chloroplast's inner membrane is a set of interconnected and stacked fluid-filled membrane sacs called thylakoids. Each stack of thylakoids is called a granum (plural = grana). The fluid enclosed by the inner membrane that surrounds the grana is called the stroma.

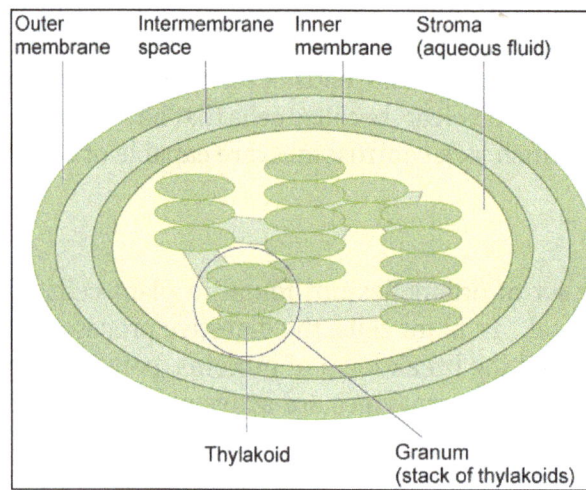

The Chloroplast Structure

In figure, the chloroplast has an outer membrane, an inner membrane, and membrane structures called thylakoids that are stacked into grana. The space inside the thylakoid membranes is called the thylakoid space. The light harvesting reactions take place in the thylakoid membranes, and the synthesis of sugar takes place in the fluid inside the inner membrane, which is called the stroma.

The chloroplasts contain a green pigment called chlorophyll, which captures the light energy that drives the reactions of photosynthesis. Like plant cells, photosynthetic protists also have chloroplasts. Some bacteria perform photosynthesis, but their chlorophyll is not relegated to an organelle.

Central Vacuole

The central vacuole plays a key role in regulating the cell's concentration of water in changing environmental conditions. When you forget to water a plant for a few days, it wilts. That's because as the water concentration in the soil becomes lower than the water concentration in the plant, water moves out of the central vacuoles and cytoplasm. As the central vacuole shrinks, it leaves the cell wall unsupported. This loss of support to the cell walls of plant cells results in the wilted appearance of the plant. The central vacuole also supports the expansion of the cell. When the central vacuole holds more water, the cell gets larger without having to invest a lot of energy in synthesizing new cytoplasm.

Prokaryotic Cells

Prokaryotic cells fall into a size range of about 1–5μm and hence can be observed clearly under microscopes. However, some prokaryotic cells may be larger than this. A prokaryotic cell consists of external and internal structures. Capsule, flagella, axial filaments, fimbriae, and pili are the external structure of the cell wall, while interior of the cell contains cytoplasm.

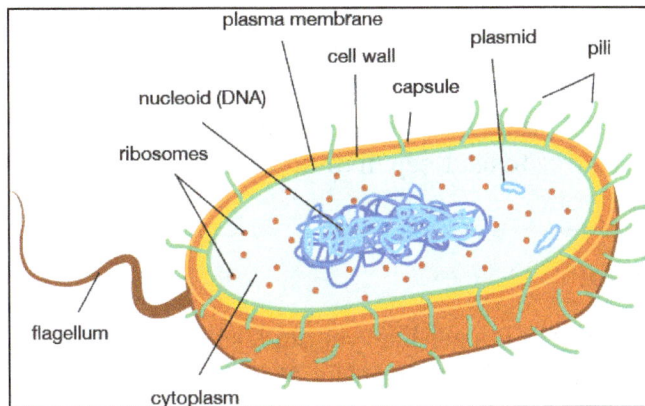

There are two major kinds of prokaryotes:

1. Bacteria (singular "bacterium")

2. Archaea (singular "archaeon")

Biologists now estimate that each human being carries nearly *20 times* more bacterial, or prokaryotic, cells in his or her body than human, or eukaryotic, cells. If that statistic overwhelms us, rest
assured that most of these bacteria are trying to help us, not hurt us.

Numerically, there are 20 times more prokaryotic cells on Earth than there are eukaryotic cells.
This is only a minimum estimate, however, because there are trillions upon trillions of bacterial
cells that are not associated with eukaryotic organisms.

In addition, all archaea are prokaryotic, too. As is the case for bacteria, it is unknown how many
archaean cells are on Earth, but the number is sure to be astronomical. In all, eukaryotic cells make
up only a very small fraction of the total number of cells on Earth.

There are four main structures shared by all prokaryotic cells, bacterial or archaean:

1. The plasma membrane

2. Cytoplasm

3. Ribosomes

4. Genetic material (DNA and RNA).

Some prokaryotic cells also have other structures like the cell wall, pili(singular "pillus"), and flagella (singular "flagellum"). Each of these structures and cellular components plays a critical role
in the growth, survival, and reproduction of prokaryotic cells.

Prokaryotic Plasma Membrane

Prokaryotic cells can have multiple plasma membranes. Prokaryotes known as "gram-negative
bacteria," for example, often have two plasma membranes with a space between them known as
the periplasm. As in all cells, the plasma membrane in prokaryotic cells is responsible for controlling what gets into and out of the cell.

A series of proteins stuck in the membrane (poor fellas) also aids prokaryotic cells in communicating with the surrounding environment. Among other things, this communication can include sending and receiving chemical signals from other bacteria and interacting with the cells of eukaryotic organisms during the process of infection. Infection, by the way, is the kind of thing that we *don't* want prokaryotes doing to us. Keep in mind that the plasma membrane is universal to all cells, prokaryotic and eukaryotic.

Prokaryotic Cytoplasm

The cytoplasm in prokaryotic cells is a gel-like, yet fluid, substance in which all of the other cellular components are suspended. Think Jell-O for cells. It is very similar to the eukaryotic cytoplasm, except that it does not contain organelles.

Recently, biologists have discovered that prokaryotic cells have a complex and functional cytoskeleton similar to that seen in eukaryotic cells. The cytoskeleton helps a prokaryotic cell to divide and to maintain its plump, round shape. As is the case in eukaryotic cells, the cytoskeleton is the framework along which particles in the cell—including proteins, ribosomes, and small rings of DNA called plasmids—move around. It's the cell's "highway system" suspended in Jell-O.

Prokaryotic Ribosomes

Prokaryotic ribosomes are smaller and have a slightly different shape and composition than those found in eukaryotic cells. Bacterial ribosomes, for instance, have about half of the amount of ribosomal RNA (rRNA) and one-third fewer ribosomal proteins (53 vs. ~83) than eukaryotic ribosomes have. Despite these differences, the function of the prokaryotic ribosome is virtually identical to the eukaryotic version. Just like in eukaryotic cells, prokaryotic ribosomes build proteins by translating messages sent from DNA.

Prokaryotic Genetic Material

All prokaryotic cells contain large quantities of genetic material in the form of DNA and RNA. Because prokaryotic cells, by definition, do *not* have a nucleus, a single large circular strand of DNA containing most of the genes needed for cell growth, survival, and reproduction is found in the cytoplasm.

This chromosomal DNA tends to look like a mess of string in the middle of the cell:

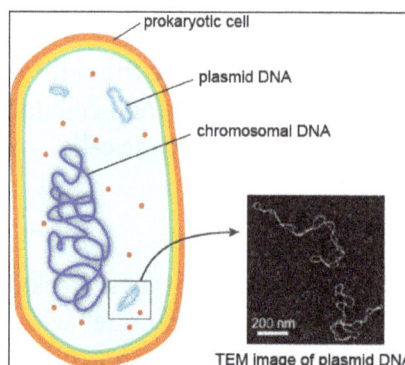

Transmission electron micrograph image

Usually, the DNA is spread throughout the entire cell, where it is readily accessible to be transcribed into messenger RNA (mRNA) that is immediately translated by ribosomes into protein. Sometimes, when biologists prepare prokaryotic cells for viewing under a microscope, the DNA will condense in one part of the cell to produce a darkened area called a nucleoid.

As in eukaryotic cells, the prokaryotic chromosome is intimately associated with special proteins involved in maintaining the chromosomal structure and regulating gene expression.

In addition to a single large piece of chromosomal DNA, many prokaryotic cells also contain small pieces of DNA called plasmids. These circular rings of DNA are replicated independently of the chromosome and can be transferred from one prokaryotic cell to another through pili, which are small projections of the cell membrane that can form physical channels with the pili of adjacent cells. The transfer of plasmids between one cell and another is often referred to as "bacterial sex."

The genes for antibiotic resistance, or the gradual ineffectiveness of antibiotics in populations, are often carried on plasmids. If these plasmids get transferred from resistant cells to nonresistant cells, bacterial infection in populations can become much harder to control. For example, it was recently learned that the superbug MRSA, or multidrug-resistant *Staphylococcus aureus*, received some of its drug-resistance genes on plasmids.

Prokaryotic cells are often viewed as "simpler" or "less complex" than eukaryotic cells. In some ways, this is true. Prokaryotic cells usually have fewer visible structures, and the structures they do have are smaller than those seen in eukaryotic cells. Do not be fooled. Just because prokaryotic cells seem "simple" does not mean that they are somehow inferior to or lower than eukaryotic cells and organisms. Making this assumption can get you into some serious trouble.

Biologists are now learning that bacteria are able to communicate and collaborate with one another on a level of complexity that rivals any communication system ever developed by humans. In addition, some archaean cells are able to thrive in environments so hostile that no eukaryotic cell would survive for more than a few seconds. You try living in a hot spring, salt lake, volcano, or even deep underground. Prokaryotic cells are also able to pull off stuff that eukaryotic cells could only dream of, in part because of their increased simplicity. Being bigger and more complex is not always better.

These cells and organisms are just as adapted to their local conditions as any eukaryote and, in that sense, are just as "evolved" as any other living organism on earth. One kind of bacterial communication, also known as quorum sensing, is where small chemical signals are used to count how many bacteria there are.

Cell Diversity

Cells show great diversity in form and functions. Because of this, it was not easy to realize that all living organisms are made up of units sharing a common basic structure. Every unit is a cell. The other major issue for the discovering of the cell was the very small size they usually show.

Cell Size

Cell size is measured in micrometers (µm). One micrometer, or micron, is one thousandth of a millimeter (10^{-3} millimeters), and one millionth of a meter (10^{-6} meters). A typical eukaryote cell is between 10 and 30 µmin size. This is true for the cells of a worm and for those of an elephant, but there are many more cells in the elephant. To be aware of how small the cells are, imagine a 1.70 meters tall person which is stretched to match the height of the Everest, which is about 8500 meters. The stretched giant cells of that person would measure only 1.3 centimeters, i.e., smaller than an euro cent coin (then, it would be a giant formed by a huge amount of euro cent coins).

However, there are eukaryote cells having uncommon dimensions. They can be very small, like sperm cells, whose head may be smaller than 4 µm in diameter, while others like the eggs of some birds and reptiles can be larger than 10 centimeters (thousands of microns) in their larger axis, but we should measure only the yolk, since the egg white is not part of the cell. An extreme example is the egg of ostriches. Some cells may have cytoplasmic extensions as large as several meters in length, such as the brain neurons of giraffes that innervate the more caudal part of the spinal cord. Smaller than eukaryote cells are prokaryote cells, which typically are around 1 to 2 µm in diameter, being *Mycoplasma* the smallest with 0.5 µm.

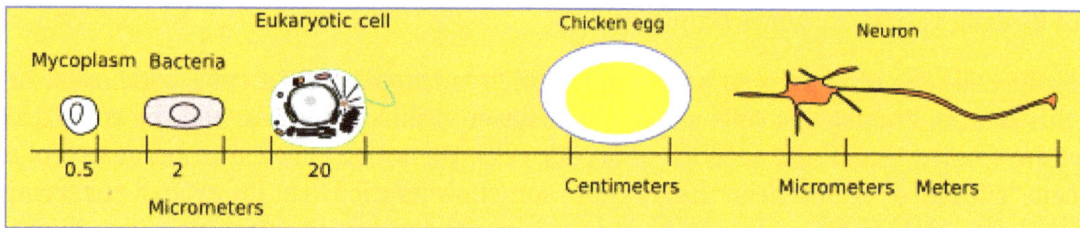

Some Examples of Cell Dimensions

Number

Most living organisms are unicellular, i.e., a single cell. Prokaryotes (bacteria and archaea) are the most abundant unicellular organisms. Unicellular eukaryote species are abundant too. Organisms that can be observed without microscopes are mostly multicellular, i.e., they are made up of many cells. Multicellular organisms are animals, plants, fungi and some algae. In general, larger multicellular organisms contain higher number of cells since they have a similar average cell size. Estimates of the total number of cells of an organism with similar size to humans may range from 10^{13} (1 followed by 13 zeros) to 10^{14} (1 followed by 14 zeros). To be aware of these numbers, it is estimated that the total number of cells in the human brain is about 86×10^9 neurons and of a mouse brain is about 15×10^9. The most abundant cells of the human body are red blood cells and glial/neuron cells of the nervous system.

Morphology

Cell morphology is typically sketched as rounded, but this is probably the most uncommon shape (except for a few types of cells). Cell morphology in animal tissues is diverse, enormously diverse. It can range from rounded to star-like, from multi-lobed to filiform. Plant cells also show a wide diversity of forms, which is determined by the cell wall, being cuboidal and columnar shapes the most common shapes.

Cell shapes:

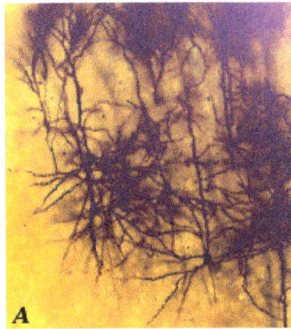

A) Neurons of the cerebral cortex

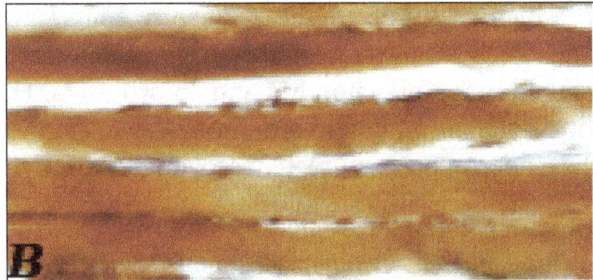

B) Skeletal muscle cells in longitudinal view

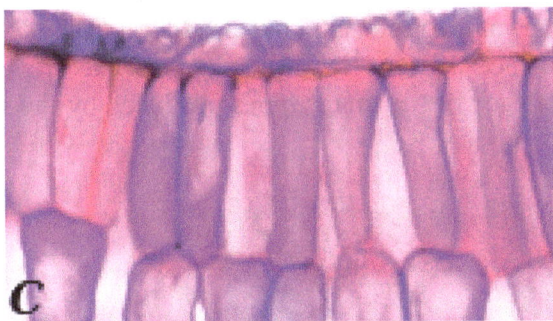

C) Cells of a leaf. Different morphologies can be observed in the parenchyma, lower part, with large and elongated cells, and in the epidermis at the upper part, with small and irregular cells located.

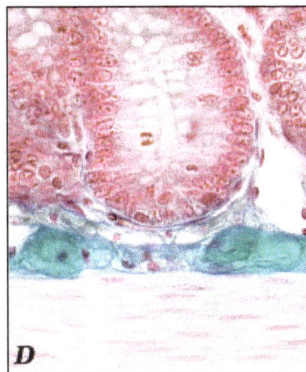

D) Different cell types of small intestine. The purple upper cells are epithelial cells, the pale elongated cells at the bottom are smooth muscle cells, and the greenish are connective tissue cells.

Molecular Make up of Cells

Cells are the basic structural and functional units of all living organisms. Cells are made up of the compounds you learnt about in the previous chapter: carbohydrates, fats, proteins, nucleic acids and water. The word 'cell' was first used by the 17th century scientist Robert Hooke to describe the small pores in a cork that he observed under a microscope. Cells are very small structures. The human body is made up of 10^{13} cells. Each of these is too small to see with the human eye and it is through the development of microscopic techniques that we have been better able to visualise and understand them.

Microscopy

Early attempts to magnify images of objects through grinding of glass lenses eventually gave rise to the earliest microscope. In 1600, Anton van Leeuwenhoek, a Dutch microbiologist used a simple microscope with only one lens to observe blood cells. He was the first scientist to describe cells and bacteria through observation under microscope. By combining two or more lenses, the magnification of the microscopes was improved, thus allowing scientists to view smaller structures.

The dissecting microscope is an optical microscope used to view images in three dimensions at low resolution. It is useful for low-level magnification of live tissue. The development of the light microscope, which uses visible light to magnify images allowed for up to 1000X magnification of objects through which scientists were able to view individual cells and internal cell structures such as the cell wall, membrane, mitochondria and chloroplasts. However, although the light microscope allowed for 1000X magnification, in order to see even smaller structures such as the internal structure of organelles, microscopes of greater resolving power (with up to 10 000X magnification) were required.

With the development of electron microscopes the microscopic detail of organelles such as mitochondria and chloroplasts became easier to observe. The Transmission Electron Microscope (TEM) was developed first, followed by the Scanning Electron Microscope (SEM). TEM is used to view extremely thin sections of material. Beams of electrons pass through the material and are focused by electromagnetic lenses. In SEM the electrons are bounced off the surface of the material and thus produce a detailed image of the external surface of the material. They produce a 3D image by picking up secondary electrons knocked off the surface with an electron collector. The image is then amplified and viewed on a screen. Examples of each of the image types produced by these microscopes are given in figures.

SEM: A natural community of bacteria growing on a single grain of sand.

SEM: These pollen grains show the characteristic depth of field of SEM micrographs.

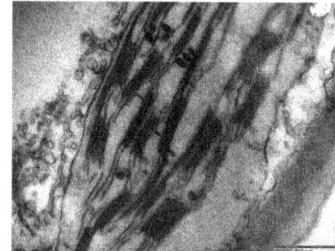

TEM: Image of chloroplast, showing thylakoid discs within a eukaryotic cell.

The apparatus most commonly used in lab microscopy exercises is a simple light microscope. Table shows an annotated diagram of a light microscope with a description of the function of each part. The main parts are described in the table that follows and the function of each part is explained.

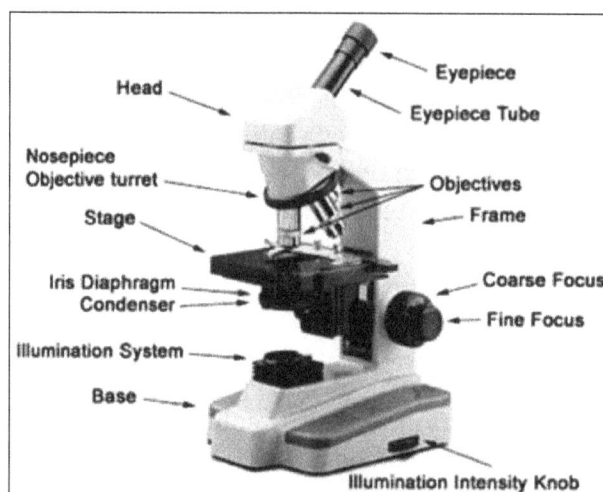

Light microscope

Table: The parts of a microscope.

Part of the microscope	Description
Ocular lens/ eyepiece	- A cylinder containing two or more lenses. - These lenses are held at the correct working distance. - The ocular lens/eyepiece helps to bring the object into focus.
Revolving nose piece	The revolving nose piece holds the objectives in place so that they can rotate and can be changed easily.
Objective	The objective magnifies the objects. There are normally three objectives present: • 4X magnification • 10X magnification • 40X magnification
Coarse adjustment screw	The coarse adjustment screw is used for the initial focus of the object. By moving the stage up and down, bringing the object closer to or further away from the objective lens.
Fine adjustment screw	The fine adjustment screw is used for the final and clear focus of the object.
Frame	A rigid structure for stability. The frame is supported by a U-shaped foot leading to the base of the microscope.
Light source / mirror	Provides a source of light so that the object can be viewed.
Diaphragm and condenser	The diaphragm and condenser control the amount of light which passes through the slide.
Stage	The microscope slide is placed here. The stage contains a clip or clips to prevent the slide from moving around. There is a hole in the stage which allows light through.

Use of Microscope Correctly

1. When handling or carrying the microscope, always do so with both hands. Grasp the arm with one hand and place the other hand under the base for support.

2. Turn the revolving nosepiece so that the lowest power objective is in position.

3. Place the microscope slide on the stage and and fasten it with the stage clips.

4. Look through the eyepiece and adjust the diaphragm for the greatest amount of light.

5. While looking at the slide on the stage from the side, turn the coarse adjustment screw so that the stage is as close to the objective lens as possible. Warning: Make sure you do not touch or damage the slide.

6. Slowly turn the coarse adjustment screw until the image comes into focus.

7. Now use the fine adjustment screw to move the stage downwards until the image is clearly visible. Never move the lens towards the slide.

8. You can readjust the light source and diaphragm for the clearest image.

9. When changing to the next objective lens use the fine adjustment screw to focus the image. Warning: Never use the coarse adjustment screw for the strongest objective lens.

10. Do not touch the glass part of the lenses with your fingers.

11. When finished, move the stage and objective as far away from each other as possible and remove the slide.

12. Disconnect the power source and cover the microscope.

13. Carry the microscope by holding it firmly by the "arm" and "base" and when walking it should be near your chest.

Differences between the light microscope and transmission electron microscope.

Property	Light Microscope	Transmission Electron Microscope
Source	Light	Beam of electrons
Resolution (how far apart two objects must be in order to be distinguished as separate)	Under optimal conditions (clean lenses, oil immersion), the resolution is 0,2, micrometres or 2 thousands of a millimeter.	Resolution of a transmission electron microscope is about 0,05 nanometres (nmnm) which is about 0,5 millionth of a millimetre. This means that a transmission electron microscope has about 10,000 times the resolving power of a light instrument.
Material (alive/ dead)	Alive or dead: Bright field or phase contrast microscopes enable viewer to observe living cells. Specimens need to be stained.	Dead: Electron microscope images are produced by passing an electron beam through tissues stained with heavy metals.
Example of microscope image	Bacterial spores as seen under light microscope.	Chlamydomonas reinhardtii, a single celled green algae, as seen under the transmission electron microscope.

Cell Biology

Cell biology is the study of cell structure and function, and it revolves around the concept that the cell is the fundamental unit of life. Focusing on the cell permits a detailed understanding of the tissues and organisms that cells compose. Some organisms have only one cell, while others are organized into cooperative groups with huge numbers of cells. On the whole, cell biology focuses on the structure and function of a cell, from the most general properties shared by all cells, to the unique, highly intricate functions particular to specialized cells.

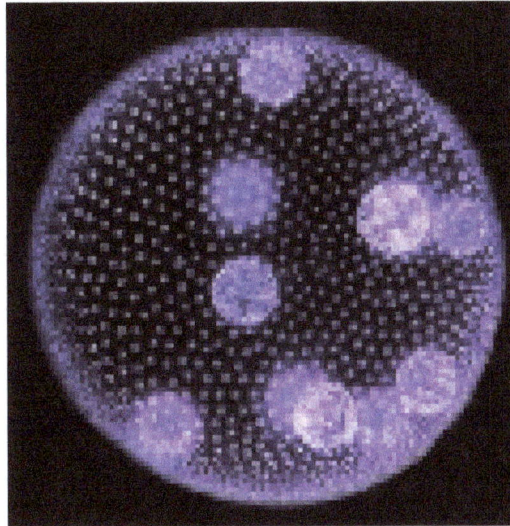

The starting point for this discipline might be considered the 1830s. Though scientists had been using microscopes for centuries, they were not always sure what they were looking at. Robert Hooke's initial observation in 1665 of plant-cell walls in slices of cork was followed shortly by Antonie van Leeuwenhoek's first descriptions of live cells with visibly moving parts. In the 1830s two scientists who were colleagues - Schleiden, looking at plant cells, and Schwann, looking first at animal cells - provided the first clearly stated definition of the cell. Their definition stated that that all living creatures, both simple and complex, are made out of one or more cells, and the cell is the structural and functional unit of life - a concept that became known as cell theory.

As microscopes and staining techniques improved over the nineteenth and twentieth centuries, scientists were able to see more and more internal detail within cells. The microscopes used by van Leeuwenhoek probably magnified specimens a few hundredfold. Today high-powered electron microscopes can magnify specimens more than a million times and can reveal the shapes of organelles at the scale of a micrometer and below. With confocal microscopy a series of images can be combined, allowing researchers to generate detailed three-dimensional representations of cells. These improved imaging techniques have helped us better understand the wonderful complexity of cells and the structures they form.

There are several main subfields within cell biology. One is the study of cell energy and the biochemical mechanisms that support cell metabolism. As cells are machines unto themselves, the focus on cell energy overlaps with the pursuit of questions of how energy first arose in original primordial cells, billions of years ago. Another subfield of cell biology concerns the genetics of the cell and its tight interconnection with the proteins controlling the release of genetic information from the nucleus to the cell cytoplasm. Yet another subfield focuses on the structure of cell components, known as subcellular compartments. Cutting across many biological disciplines is the additional subfield of cell biology, concerned with cell communication and signaling, concentrating on the messages that cells give to and receive from other cells and themselves. And finally, there is the subfield primarily concerned with the cell cycle, the rotation of phases beginning and ending with cell division and focused on different periods of growth and DNA replication. Many cell biologists dwell at the intersection of two or more of these subfields as our ability to analyze cells in more complex ways expands.

In line with continually increasing interdisciplinary study, the recent emergence of systems biology has affected many biological disciplines; it is a methodology that encourages the analysis of living systems within the context of other systems. In the field of cell biology, systems biology has enabled the asking and answering of more complex questions, such as the interrelationships of gene regulatory networks, evolutionary relationships between genomes, and the interactions between intracellular signaling networks. Ultimately, the broader a lens we take on our discoveries in cell biology, the more likely we can decipher the complexities of all living systems, large and small.

References

- Cell: biologydictionary.net, Retrieved 1 May, 2019

- Eukaryotic-cells, boundless-biology: lumenlearning.com, Retrieved 4 Feb, 2019

- Prokaryotic-cells, biology-cells: shmoop.com, Retrieved 4 January, 2019

- Diversidad, english: webs.uvigo.es, Retrieved 13 Febuary, 2019

- Cells-the-basic-units-of-life, lifesciences, science, read: siyavula.com, Retrieved 17 March, 2019

- Cell-biology: nature.com, Retrieved 1 May, 2019

Chapter 2

Composition of Cells

Cells are composed of various macromolecules such as nucleic acid, amino acids and proteins. Besides these, the extracellular macromolecules which are present outside the cells provide biochemical and structural support to the cells around them. The network of these molecules is known as the extracellular matrix. This chapter has been carefully written to provide an easy understanding of these macromolecules in cells.

Chemical Composition of Living Cells

All living organisms, from microbes to mammals, are composed of chemical substances from both the inorganic and organic world, that appear in roughly the same proportions, and perform the same general tasks. Hydrogen, oxygen, nitrogen, carbon, phosphorus, and sulfur normally make up more than 99% of the mass of living cells, and when combined in various ways, form virtually all known organic biomolecules. They are initially utilized in the synthesis of a small number of building blocks that are, in turn, used in the construction of a vast array of vital macromolecules.

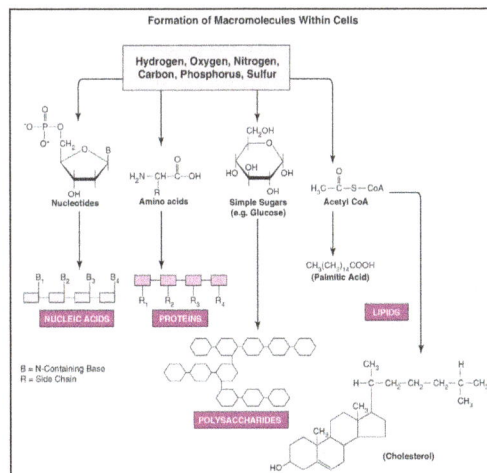

Formation of Macromolecules within Cells

There are four general classes of macromolecules within living cells: nucleic acids, proteins, polysaccharides, and lipids. These compounds, which have molecular weights ranging from 1×10^3 to 1×10^6, are created through polymerization of building blocks that have molecular weights in the range of 50 to 150. Although subtle differences do exist between cells (e.g., erythrocyte, liver, muscle or fat cell), they all generally contain a greater variety of proteins than any other type of macromolecule, with about 50% of the solid matter of the cell being protein (15% on a wet weight basis). Cells generally contain many more protein molecules than DNA molecules, yet DNA is typically the largest biomolecule in the cell. About 99% of cellular molecules are water molecules, with water normally accounting for approximately 70% of the total wet-weight of the cell. Although water is obviously important to the vitality of all living cells, the bulk of our attention is usually focused on the other 1% of biomolecules.

Each E. coli, and similar bacterium, contains a single chromosome; therefore, it has only one unique DNA molecule. Mammals, however, contain more chromosomes, and thus have different DNA molecules in the nucleus.

Nucleic Acids

Nucleic acids are nucleotide polymers that store and transmit genetic information. Only 4 different nucleotides are used in nucleic acid biosynthesis. Genetic information contained in nucleic acids is stored and replicated in chromosomes, which contain genes. A chromosome is a deoxyribonucleic acid (DNA) molecule, and genes are segments of intact DNA. The total number of genes in any given mammalian cell may total several thousand. When a cell replicates itself, identical copies of DNA molecules are produced, therefore the hereditary line of descent is conserved, and the genetic information carried on DNA is available to direct the occurrence of virtually all chemical reactions within the cell. The bulk of genetic information carried on DNA provides instructions for the assembly of virtually every protein molecule within the cell. The flow of information from nucleic acids to protein is commonly represented as DNA —> messenger ribonucleic acid (mRNA) —> transfer RNA (tRNA) —> ribosomal RNA (rRNA) —> protein, which indicates that the nucleotide sequence in a gene of DNA specifies the assembly of a nucleotide sequence in an mRNA molecule, which in turn directs the assembly of the amino acid sequence in protein through a tRNA and rRNA molecules.

Proteins

Proteins are amino acid polymers responsible for implementing instructions contained within the genetic code. Twenty different amino acids are used to synthesize proteins about half are formed as metabolic intermediates, while the remainder must be provided through the diet. The latter group is referred to as "essential" amino acids. Each protein formed in the body, unique in its own structure and function, participates in processes that characterize the individuality of cells, tissues, organs, and organ systems. A typical cell contains thousands of different proteins, each with a different function, and many serve as enzymes that catalyze (or speed) reactions. Virtually every reaction in a living cell requires an enzyme. Other proteins transport different compounds either outside or inside cells {e.g., lipoproteins and transferrin (an iron-binding protein) in plasma, or bilirubin-binding proteins in liver cells}; some act as storage proteins (e.g., myoglobin binds and stores O_2 in muscle cells); others as defense proteins in blood or on the surface of cells (e.g., clotting proteins and immunoglobulins); others as contractile proteins (e.g., the actin, myosin and troponin of skeletal muscle fibers); and others are merely structural in nature (e.g., collagen and elastin). Proteins, unlike glycogen and triglyceride, are usually not synthesized and stored as nonfunctional entities.

Polysaccharides

Polysaccharides are polymers of simple sugars (i.e., monosaccharides). Some polysaccharides are homogeneous polymers that contain only one kind of sugar (e.g., glycogen), while others are complex heterogenous polymers that contain 8-10 types of sugars. In contrast to heterogenous polymers (e.g., proteins, nucleic acids, and some polysaccharides), homogenous polymers are considered to be "noninformational". Polysaccharides, therefore, can occur as functional and structural components of cells (e.g., glycoproteins and glycolipids), or merely as noninformational storage forms of energy (e.g., glycogen). The 8-10 monosaccharides that become the building blocks for heterogenous polysaccharides can be synthesized from glucose, or formed from other metabolic intermediates.

Lipids

Lipids are naturally occurring, nonpolar substances that are mostly insoluble in water (with the exceptions being the short-chain volatile fatty acids and ketone bodies), yet soluble in nonpolar solvents (like chloroform and ether). They serve as membrane components (cholesterol, glycolipids and phospholipids), storage forms of energy (triglycerides), precursors to other important biomolecules (fatty acids), insulation barriers (neutral fat stores), protective coatings to prevent infection and excessive gain or loss of water, and some vitamins (A, D, E, and K) and hormones (steroid hormones). Major classes of lipids are the saturated and unsaturated fatty acids (short, medium, and long-chain), triglycerides, lipoproteins {i.e., chylomicrons (CMs), very low density (VLDL), low density (LDL), intermediate density (IDL), and high density lipoproteins (HDL)}, phospholipids and glycolipids, steroids (cholesterol, progesterone, etc.), and eicosanoids (prostaglandins, thromboxanes, and leukotrienes). All lipids can be synthesized from acetyl-CoA, which in turn can be generated from numerous different sources, including carbohydrates, amino acids, short-chain volatile fatty acids (e.g., acetate), ketone bodies, and fatty acids. Simple lipids include only those that are esters of fatty acids and an alcohol (e.g., mono-, di- and triglycerides). Compound lipids include various materials that contain other substances in addition to an alcohol and fatty acid (e.g., phosphoacylglycerols, sphingomyelins, and cerebrosides), and derived lipids include those that cannot be neatly classified into either of the above (e.g., steroids, eicosanoids, and the fat-soluble vitamins).

Although the study of physiological chemistry emphasizes organic molecules, the inorganic elements (sometimes subdivided into macrominerals, trace elements, and ultra-trace elements), are also important. Several are "essential" nutrients, and therefore like certain amino acids and unsaturated fatty acids, must be supplied in the diet. Inorganic elements are typically present in cells as ionic forms, existing as either free ions or complexed with organic molecules. Many "trace elements" are known to be essential for life, health, and reproduction, and have well-established actions (e.g., cofactors for enzymes, sites for binding of oxygen (in transport), and structural components of nonenzymatic macromolecules). Some investigators have speculated that perhaps all of the elements on the periodic chart will someday by shown to exhibit physiologic roles in mammalian life.

Because life depends upon chemical reactions, and because most all diseases in animals are manifestations of abnormalities in biomolecules, chemical reactions, or biochemical pathways, physiological chemistry has become the language of all basic medical sciences. A fundamental understanding of this science is therefore needed not only to help illuminate the origin of disease, but also to help formulate appropriate therapies. The chapters which follow were designed, therefore, to assist the reader in developing a basic rational approach to the practice of veterinary medicine.

Nucleic Acids

Nucleic acids are molecules that allow organisms to transfer genetic information from one generation to the next. These macromolecules store the genetic information that determines traits and makes protein synthesis possible.

Structure of Nucleic Acids

Nucleic acids are the most important macromolecules for the continuity of life. They carry the genetic blueprint of a cell and carry instructions for the functioning of the cell. The two main types of nucleic acids are deoxyribonucleic acid (DNA) and ribonucleic acid (RNA). DNA is the genetic material found in all living organisms, ranging from single-celled bacteria to multicellular mammals. It is found in the nucleus of eukaryotes and in the organelles, chloroplasts, and mitochondria. In prokaryotes, the DNA is not enclosed in a membranous envelope.

The entire genetic content of a cell is known as its genome, and the study of genomes is genomics. In eukaryotic cells but not in prokaryotes, DNA forms a complex with histone proteins to form chromatin, the substance of eukaryotic chromosomes. A chromosome may contain tens of thousands of genes. Many genes contain the information to make protein products; other genes code for RNA products. DNA controls all of the cellular activities by turning the genes "on" or "off."

The other type of nucleic acid, RNA, is mostly involved in protein synthesis. The DNA molecules never leave the nucleus but instead use an intermediary to communicate with the rest of the cell. This intermediary is the messenger RNA (mRNA). Other types of RNA—like rRNA, tRNA, and microRNA—are involved in protein synthesis and its regulation. DNA and RNA are made up of monomers known as nucleotides. The nucleotides combine with each other to form a polynucleotide, DNA or RNA. Each nucleotide is made up of three components: a nitrogenous base, a pentose (five-carbon) sugar, and a phosphate group. Each nitrogenous base in a nucleotide is attached to a sugar molecule, which is attached to one or more phosphate groups.

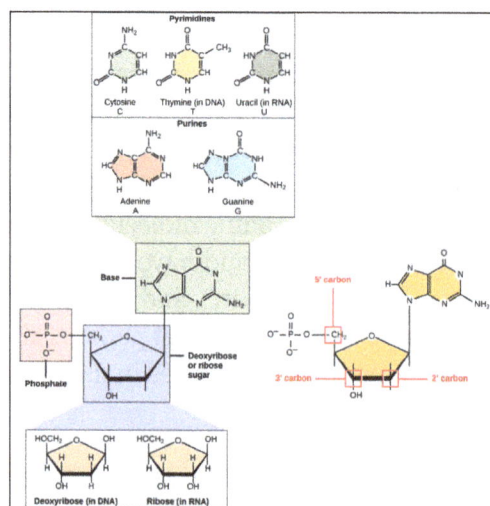

In figure, a nucleotide is made up of three components: a nitrogenous base, a pentose sugar, and one or more phosphate groups. Carbon residues in the pentose are numbered 1' through 5' (the prime distinguishes these residues from those in the base, which are numbered without using a prime notation). The base is attached to the 1' position of the ribose, and the phosphate is attached to the 5' position. When a polynucleotide is formed, the 5' phosphate of the incoming nucleotide attaches to the 3' hydroxyl group at the end of the growing chain. Two types of pentose are found in nucleotides, deoxyribose (found in DNA) and ribose (found in RNA). Deoxyribose is similar in structure to ribose, but it has an H instead of an OH at the 2' position. Bases can be divided into two categories: purines and pyrimidines. Purines have a double ring structure, and pyrimidines have a single ring.

The nitrogenous bases, important components of nucleotides, are organic molecules and are so named because they contain carbon and nitrogen. They are bases because they contain an amino group that has the potential of binding an extra hydrogen, and thus, decreases the hydrogen ion concentration in its environment, making it more basic. Each nucleotide in DNA contains one of four possible nitrogenous bases: adenine (A), guanine (G) cytosine (C), and thymine (T). RNA nucleotides also contain one of four possible bases: adenine, guanine, cytosine, and uracil (U) rather than thymine.

Adenine and guanine are classified as purines. The primary structure of a purine is two carbon-nitrogen rings. Cytosine, thymine, and uracil are classified as pyrimidines which have a single carbon-nitrogen ring as their primary structure (Figure 1). Each of these basic carbon-nitrogen rings has different functional groups attached to it. In molecular biology shorthand, the nitrogenous bases are simply known by their symbols A, T, G, C, and U. DNA contains A, T, G, and C whereas RNA contains A, U, G, and C.

The pentose sugar in DNA is deoxyribose, and in RNA, the sugar is ribose. The difference between the sugars is the presence of the hydroxyl group on the second carbon of the ribose and hydrogen on the second carbon of the deoxyribose. The carbon atoms of the sugar molecule are numbered as 1', 2', 3', 4', and 5' (1' is read as "one prime"). The phosphate residue is attached to the hydroxyl group of the 5' carbon of one sugar and the hydroxyl group of the 3' carbon of the sugar of the next nucleotide, which forms a 5'–3' phosphodiester linkage. The phosphodiester linkage is not formed by simple dehydration reaction like the other linkages connecting monomers in macromolecules: its formation involves the removal of two phosphate groups. A polynucleotide may have thousands of such phosphodiester linkages.

DNA Double-helix Structure

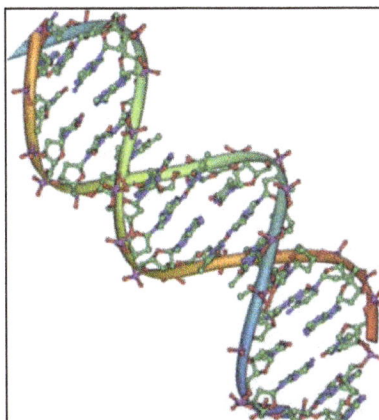

DNA is an antiparallel double helix. The phosphate backbone (the curvy lines) is on the outside, and the bases are on the inside. Each base interacts with a base from the opposing strand.

DNA has a double-helix structure. The sugar and phosphate lie on the outside of the helix, forming the backbone of the DNA. The nitrogenous bases are stacked in the interior, like the steps of a staircase, in pairs; the pairs are bound to each other by hydrogen bonds. Every base pair in the double helix is separated from the next base pair by 0.34 nm.

The two strands of the helix run in opposite directions, meaning that the 5′ carbon end of one strand will face the 3′ carbon end of its matching strand. (This is referred to as antiparallel orientation and is important to DNA replication and in many nucleic acid interactions.)

Only certain types of base pairing are allowed. For example, a certain purine can only pair with a certain pyrimidine. This means A can pair with T, and G can pair with C, as shown in figure. This is known as the base complementary rule. In other words, the DNA strands are complementary to each other. If the sequence of one strand is AATTGGCC, the complementary strand would have the sequence TTAACCGG. During DNA replication, each strand is copied, resulting in a daughter DNA double helix containing one parental DNA strand and a newly synthesized strand.

RNA

Ribonucleic acid, or RNA, is mainly involved in the process of protein synthesis under the direction of DNA. RNA is usually single-stranded and is made of ribonucleotides that are linked by phosphodiester bonds. A ribonucleotide in the RNA chain contains ribose (the pentose sugar), one of the four nitrogenous bases (A, U, G, and C), and the phosphate group.

- RNA: Messenger RNA (mRNA), ribosomal RNA (rRNA), transfer RNA (tRNA), and microRNA (miRNA). The first, mRNA, carries the message from DNA, which controls all of the cellular activities in a cell. If a cell requires a certain protein to be synthesized, the gene for this product is turned "on" and the messenger RNA is synthesized in the nucleus. The RNA base sequence is complementary to the coding sequence of the DNA from which it has been copied. However, in RNA, the base T is absent and U is present instead. If the DNA strand has a sequence AATTGCGC, the sequence of the complementary RNA is UUAACGCG. In the cytoplasm, the mRNA interacts with ribosomes and other cellular machinery.

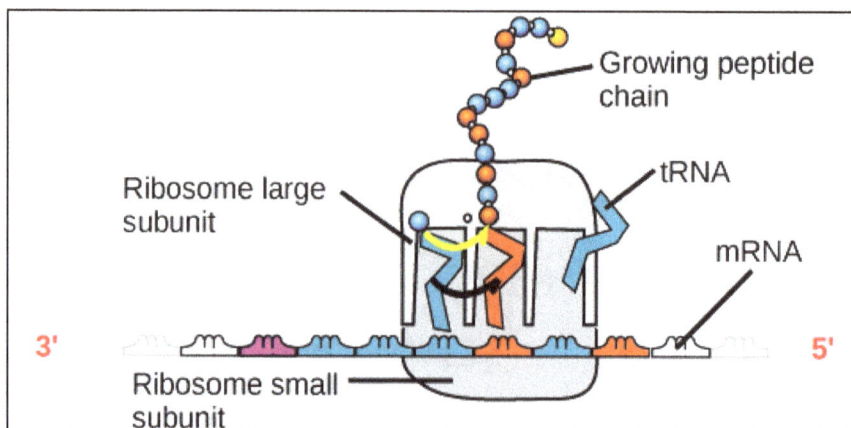

- A ribosome has two parts: A large subunit and a small subunit. The mRNA sits in between the two subunits. A tRNA molecule recognizes a codon on the mRNA, binds to it by complementary base pairing, and adds the correct amino acid to the growing peptide chain.

The mRNA is read in sets of three bases known as codons. Each codon codes for a single amino acid. In this way, the mRNA is read and the protein product is made. Ribosomal RNA (rRNA) is a major constituent of ribosomes on which the mRNA binds. The rRNA ensures the proper alignment of the mRNA and the ribosomes; the rRNA of the ribosome also has an enzymatic activity (peptidyl transferase) and catalyzes the formation of the peptide bonds between two aligned amino acids. Transfer RNA (tRNA) is one of the smallest of the four types of RNA, usually 70–90 nucleotides long. It carries the correct amino acid to the site of protein synthesis. It is the base pairing between the tRNA and mRNA that allows for the correct amino acid to be inserted in the polypeptide chain. microRNAs are the smallest RNA molecules and their role involves the regulation of gene expression by interfering with the expression of certain mRNA messages.

DNA versus RNA

While DNA and RNA are similar, they have very distinct differences. Table summarizes features of DNA and RNA.

Table: Features of DNA and RNA.

	DNA	RNA
Function	Carries genetic information	Involved in protein synthesis
Location	Remains in the nucleus	Leaves the nucleus
Structure	DNA is double-stranded "ladder": sugar-phosphate backbone, with base rungs.	Usually single-stranded
Sugar	Deoxyribose	Ribose
Pyrimidines	Cytosine, thymine	Cytosine, uracil
Purines	Adenine, guanine	Adenine, guanine

One other difference bears mention. There is only one type of DNA. DNA is the heritable information that is passed along to each generation of cells; its strands can be "unzipped" with small amount of energy when DNA needs to replicate, and DNA is transcribed into RNA. There are mutliple types of RNA: Messenger RNA is a temporary molecule that transports the information necessary to make a protein from the nucleus (where the DNA remains) to the cytoplasm, where the ribosomes are. Other kinds of RNA include ribosomal RNA (rRNA), transfer RNA (tRNA), small nuclear RNA (snRNA), and microRNA.

Even though the RNA is single stranded, most RNA types show extensive intra-molecular base pairing between complementary sequences, creating a predictable three-dimensional structure essential for their function.

Information flow in an organism takes place from DNA to RNA to protein. DNA dictates the structure of mRNA in a process known as transcription, and RNA dictates the structure of protein in a process known as translation. This is known as the Central Dogma of Life, which holds true for all organisms; however, exceptions to the rule occur in connection with viral infections.

Carbon Compounds in Cells

Carbon has unique quantum properties that make it the backbone of all organic compounds. It cycles through the atmosphere and the biomass of living organisms through various biochemical pathways.

Properties of Organic Compounds

1. An organic compound consists of carbon and one or more additional elements, covalently bonded to one another.

2. Effects of Carbon's Bonding Behavior:

- Oxygen, hydrogen, and carbon are the most abundant elements in living matter.

- Much of the H and O are linked as water.

- Carbon can share pairs of electrons with as many as four other atoms to form organic molecules of several configurations.

- A carbon atom can rotate freely around a single covalent bond.

- A double covalent bond restricts rotation.

- Such interactions help give rise to the three-dimensional shapes and functions of biological molecules.

3. Hydrocarbons and Functional Groups:

- In hydrocarbons, only hydrogen atoms are attached to the carbon backbone; these molecules are quite stable.

- Functional groups are atoms or groups of atoms covalently bonded to a carbon backbone; they convey distinct properties, such as solubility and chemical reactivity, to the complete molecule.

Cells use Organic Compounds

1. Five Classes of Reactions:

- Enzymes are a special class of proteins that mediate five categories of reactions.

  ○ Functional-group transfer from one molecule to another,

  ○ Electron transfer –stripped from one molecule and given to another,

  ○ Rearrangement of internal bonds converts one type of organic molecule to another,

  ○ Condensation of two molecules into one,

  ○ Cleavage of one molecule into two.

- In a condensation reaction, one molecule is stripped of its H^+, another is stripped of its OH^-; and then the two molecule fragments join to form a new compound and the H^+ and OH^- form water.

- Hydrolysis is the reverse: one molecule is split by the addition of H^+ and OH^- (from water) to the components.

Carbohydrates

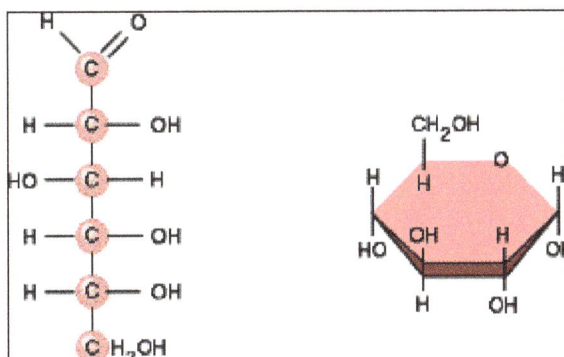

1. The Simple Sugars:

- Monosaccharides – one sugar unit – are the simplest carbohydrates.

- They are characterized by solubility in water, sweet taste, and several ^-OH groups.

- Ribose and deoxyribose (five-carbon backbones) are building blocks for nucleic acids.

- Glucose and fructose (six-carbon backbones) are used in assembling larger carbohydrates.

- Other important molecules derived from sugar monomers include glycerol and vitamin C.

2. Short-Chain Carbohydrates:

- An oligosaccharide is a short chain of two or more sugar monomers.

- Disaccharides—two sugar units—are the simplest.

  ◦ Lactose (glucose + galactose) is present in milk.

  ◦ Sucrose (glucose + fructose) is a transport form of sugar used by plants and harvested by humans for use in food.

  ◦ Maltose (two glucose units) is present in germinating seeds.

- Oligosaccharides with three or more sugar monomers are attached as short side chains to proteins where they participate in membrane function.

3. Complex Carbohydrates:

- A polysaccharide is a straight or branched chain of hundreds or thousands of sugar monomers.

- Starch is a plant storage form of energy, arranged as unbranched coiled chains, easily hydrolyzed to glucose units.

- Cellulose is a fiber-like structural material–tough, insoluble–used in plant cell walls.

- Glycogen is a highly branched chain used by animals to store energy in muscles and liver.

- Chitin is a specialized polysaccharide with nitrogen attached to the glucose units; it is used as a structural material in arthropod exoskeletons and fungal cell walls.

Lipids

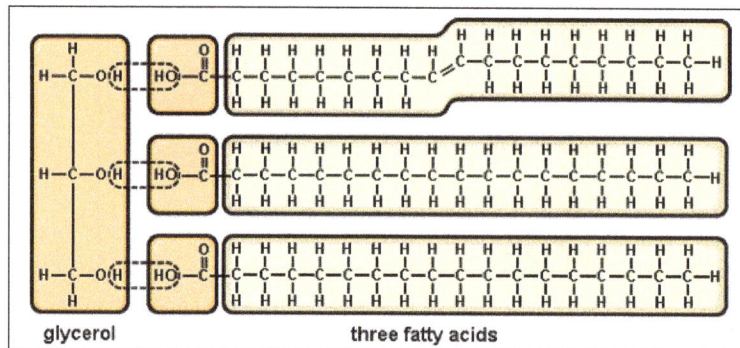

glycerol   three fatty acids

Lipids are greasy or oily non-polar compounds that function in energy storage, membrane structure, and coatings.

1. Fatty Acids:

- A fatty acid is a long chain of mostly carbon and hydrogen atoms with a –COOH group at one end.

- When they are part of complex lipids, the fatty acids resemble long, flexible tails:

  ◦ Unsaturated fats are liquids (oils) at room temperature because one or more double bonds between the carbons in the fatty acids permits "kinks" in the tails.

  ◦ Saturated fats (triglycerides) have only single C–C bonds in their fatty acid tails and are solids at room temperature.

2. Triglycerides (Neutral Fats):

- These are formed by the attachment of one (mono-), two (di-), or three (tri-) fatty acids to a glycerol.

- They are a rich source of energy, yielding more than twice the energy per weight basis as carbohydrates.

3. Phospholipids:

- They are formed by attachment of two fatty acids plus a phosphate group to a glycerol.

- They are the main structural material of membranes where they arrange in bilayers.

4. Sterols and Their Derivatives:

- Sterols have a backbone of four carbon rings but no fatty acid tails.

- Cholesterol is a component of cell membranes in animals and can be modified to form sex hormones (testosterone and estrogen) and vitamin D.

5. Waxes:

- They are formed by attachment of long-chain fatty acids to long-chain alcohols or carbon rings.

- They serve as coatings for plant parts and as animal coverings.

Proteins

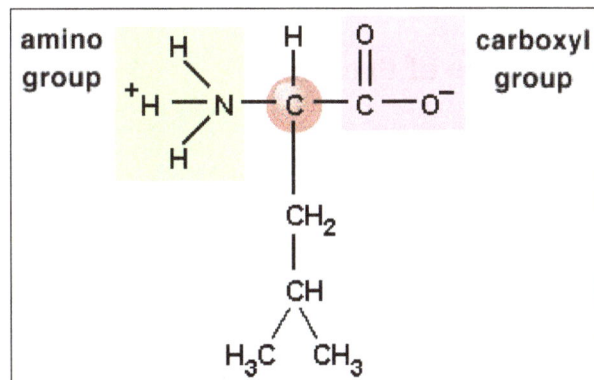

Proteins are composed of amino acids. They function as enzymes, in cell movements, as storage and transport agents, hormones, antibodies, and structural material.

1. Structure of Amino Acids:

- Amino acids are small organic molecules with an amino group, a carboxyl group, and one of twenty varying R groups.

- All of the parts of an amino acid molecule are covalently bonded to a central carbon atom.

2. Primary Structure of Proteins:

- Primary structure is defined as ordered sequences of amino acids each linked together by peptide bonds to form polypeptide chains.

- There are 20 kinds of amino acids available in nature.

- The sequence of the amino acids is determined by DNA and is unique for each kind of protein.

  - Fibrous proteins have polypeptide chains organized as strands or sheets; they contribute to the shape, internal organization, and movement of cells.

  - Globular proteins, including most enzymes, have their chains folded into compact, rounded shapes.

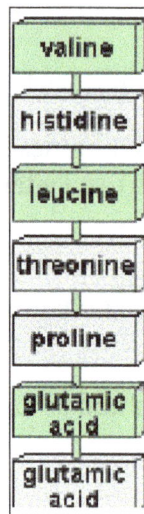

Three-dimensional Structure of Proteins

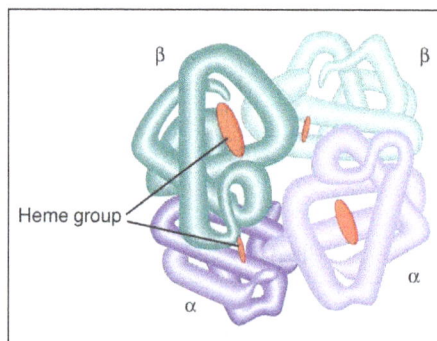

Three-dimensional structure is determined by how amino acid
sequences present their atoms for hydrogen bonding.

1. Protein Structure:

- Secondary structure refers to the helical coil (as in hemoglobin) or sheet-like array (as in silk) that results from hydrogen bonding of side groups on the amino acid chains.

- The peptide bonds between the amino acids of primary structure allow slight bending to permit secondary structure.

2. Protein Structure:

- Tertiary structure is the result of folding due to interactions among R groups along the polypeptide chain.

- The result is a more compact, globular shape in complex proteins.

3. Protein Structure:

- Quaternary structure describes the complexing of two of more polypeptide chains.

- Hemoglobin is a good example of four interacting chains that form globular proteins; keratin and collagen are complex fibrous proteins.

4. Glycoproteins and Lipoproteins:

- Lipoproteins have both lipid and protein components; they transport fats and cholesterol in the blood.

- Glycoproteins consist of oligosaccharides covalently bonded to proteins; they are abundant on the exterior of animal cells, as cell products, and in the blood.

5. Structural Changes by Denaturation:

- High temperatures or changes in pH can cause a loss of a protein's normal three-dimensional shape (denaturation).

- Normal functioning is lost upon denaturation, which is often irreversible (for example, a cooked egg).

Nucleotides and the Nucleic Acids

Nucleotides and nucleic acids are the molecular source of what we know as life. They perpetuate themselves by a self-replicating process that began on our planet about 3.5 billion years ago. Since they have achieved relatively complex systems (cells, organelles, organisms, etc.) for preserving themselves, demonstrating the ultimate example of teleology. Everything from our self-sustaining metabolic pathways to our social behavior participates intimately in insuring their own molecular integrity.

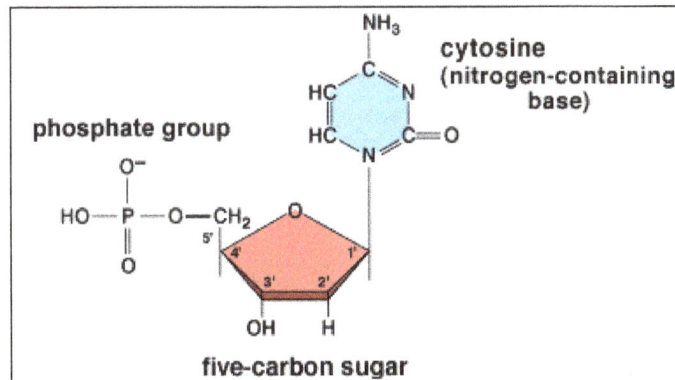

1. Nucleotides with Roles in Metabolism:

- Each nucleotide consists of a five-carbon sugar (ribose or deoxyribose), a nitrogen-containing base, and a phosphate group.

- Adenosine phosphates are chemical messengers (cAMP) or energy carriers (ATP).

- Nucleotide coenzymes transport hydrogen atoms and electrons (examples: NAD^+ and FAD).

2. Nucleic Acids - DNA and RNA:

- Nucleic acids are polymers of nucleotides.

  - Four different kinds of nucleotides are strung together to form large single or double-stranded molecules.

- ◦ Each strand's backbone consists of joined sugars and phosphates with nucleotide bases projecting toward the interior.

- • The two most important nucleic acids are DNA and RNA.

  - ◦ DNA is a double-stranded helix that stores encoded hereditary instructions.

  - ◦ RNA is single stranded and functions in translating the code to build proteins.

Amino Acids

Amino acids are used in every cell of your body to build the proteins you need to survive. All organisms need some proteins, whether they are used in muscles or as simple structures in the cell membrane. Even though all organisms have differences, they still have one thing in common: the need for basic chemical building blocks.

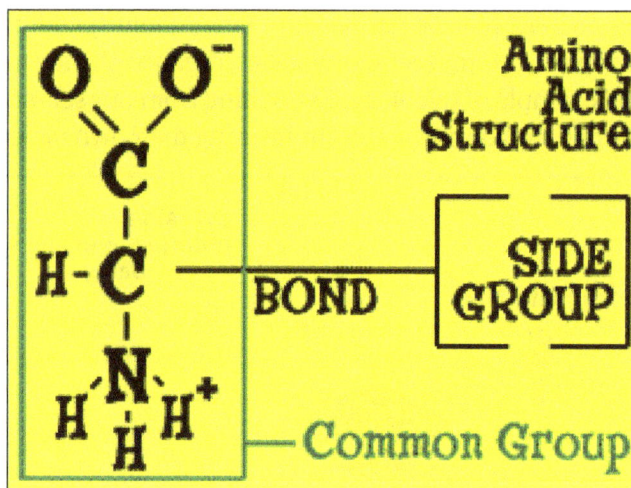

Amino acids have a two-carbon bond. One of the carbons is part of a group called the carboxyl group (COO-). A carboxyl group is made up of one carbon (C) and two oxygen (O) atoms. That carboxyl group has a negative charge, since it is a carboxylic acid (-COOH) that has lost its hydrogen (H) atom. What is left - the carboxyl group - is called a conjugate base. The second carbon is connected to the amino group. Amino means there is an NH_2 group bonded to the carbon atom. In the image, you see a "+" and a "-". Those positive and negative signs are there because, in amino acids, one hydrogen atom moves to the other end of the molecule. An extra "H" gives you a positive charge.

Making Chains

Even though scientists have discovered over 50 amino acids, only 20 are used to make something called proteins in your body. Of those twenty, nine are defined as essential. The other eleven can be synthesized by an adult body. Thousands of combinations of those twenty are used to make all of the proteins in your body. Amino acids bond together to make long chains. Those long chains of amino acids are also called proteins.

- Essential Amino Acids: Histidine, Isoleucine, Leucine, Lysine, Methionine, Phenylalanine, Threonine, Tryptophan, and Valine.

- Nonessential Amino Acids: Alanine, Asparagine, Aspartic Acid, Glutamic Acid.

- Conditional Amino Acids: Arginine (essential in children, not in adults), Cysteine, Glutamine, Glycine, Proline, Serine, and Tyrosine.

Side Groups

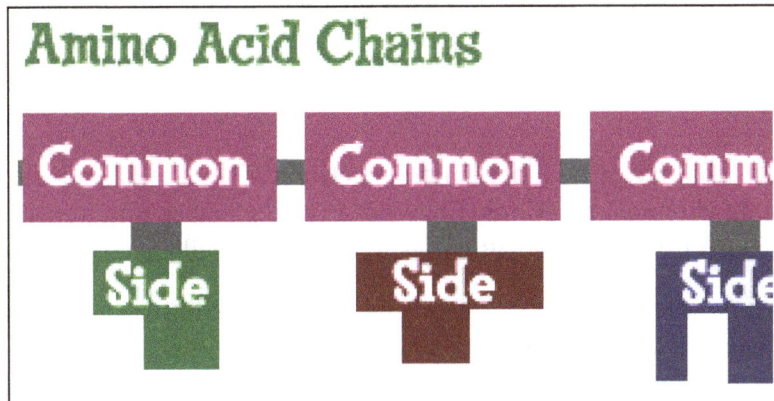

The side groups are what make each amino acid different from the others. Of the 20 side groups used to make proteins, there are two main groups: polar and non-polar. These names refer to the way the side groups, sometimes called "R" groups, interact with the environment. Polar amino acids like to adjust themselves in a certain direction. Non-polar amino acids don't really care what's going on around them. The polar and nonpolar chemical traits allow amino acids to point towards water (hydrophilic) or away from water (hydrophobic). The growing chains can then begin to twist and turn when they are being synthesized.

Proteins

Proteins are very important molecules in our cells and are essential for all living organisms. By weight, proteins are collectively the major component of the dry weight of cells and are involved in virtually all cell functions.

Each protein within the body has a specific function, from cellular support to cell signaling and cellular locomotion. In total, there are seven types of proteins, including antibodies, enzymes, and some types of hormones, such as insulin.

While proteins have many diverse functions, all are typically constructed from one set of 20 amino acids. The structure of a protein may be globular or fibrous, and the design helps each protein with their particular function.

In all, proteins are absolutely fascinating and a complex subject. Let's explore the basics of these essential molecules and discover what they do for us.

Antibodies

Antibodies are specialized proteins involved in defending the body from antigens (foreign invaders). They can travel through the bloodstream and are utilized by the immune system to identify and defend against bacteria, viruses, and other foreign intruders. One way antibodies counteract antigens is by immobilizing them so they can be destroyed by white blood cells.

Contractile Proteins

Contractile proteins are responsible for muscle contraction and movement. Examples of these proteins include actin and myosin.

Enzymes

Enzymes are proteins that facilitate biochemical reactions. They are often referred to as catalysts because they speed up chemical reactions. Enzymes include lactase and pepsin, which you might hear of often when learning about specialty diets or digestive medical conditions.

Lactase breaks down the sugar lactose found in milk. Pepsin is a digestive enzyme that works in the stomach to break down proteins in food.

Other examples of digestive enzymes are the enzymes present in saliva. Salivary amylase, salivary kallikrein, and lingual lipase all perform important biological functions. Salivary amylase is the primary enzyme found in saliva and it helps to break down starch into sugar.

Hormonal Proteins

Hormonal proteins are messenger proteins which help to coordinate certain bodily activities. Examples include insulin, oxytocin, and somatotropin.

Insulin regulates glucose metabolism by controlling the blood-sugar concentration. Oxytocin stimulates contractions during childbirth. Somatotropin is a growth hormone that stimulates protein production in muscle cells.

Structural Proteins

Structural proteins are fibrous and stringy and because of this formation, they provide support for various body parts. Examples include keratin, collagen, and elastin.

Keratins strengthen protective coverings such as skin, hair, quills, feathers, horns, and beaks. Collagens and elastin provide support for connective tissues such as tendons and ligaments.

Storage Proteins

Storage proteins store amino acids for the body to use later. Examples include ovalbumin, which is found in egg whites, and casein, a milk-based protein. Ferritin is another protein that stores iron in the transport protein, hemoglobin.

Transport Proteins

Transport proteins are carrier proteins which move molecules from one place to another around the body. Hemoglobin is one of these and is responsible for transporting oxygen through the blood via red blood cells. Cytochromes are another that operates in the electron transport chain as electron carrier proteins.

Amino Acids and Polypeptide Chains

Amino acids are the building blocks of all proteins, no matter their function. Most amino acids follow a particular structural property in which a carbon (the alpha carbon) is bonded to four different groups:

- A hydrogen atom (H)

- A Carboxyl group (-COOH)

- An Amino group (-NH$_2$)

- A "variable" group

Of the 20 amino acids that typically make up proteins, the "variable" group determines the differences among the amino acids. All amino acids have the hydrogen atom, carboxyl group, and amino group bonds.

Amino acids are joined together through dehydration synthesis to form a peptide bond. When a number of amino acids are linked together by peptide bonds, a polypeptide chain is formed. One or more polypeptide chains twisted into a 3-D shape forms a protein.

Protein Structure

We can divide the structure of protein molecules into two general classes: globular proteins and fibrous proteins. Globular proteins are generally compact, soluble, and spherical in shape. Fibrous proteins are typically elongated and insoluble. Globular and fibrous proteins may exhibit one or more types of protein structure.

There are four levels of protein structure: primary, secondary, tertiary, and quaternary. These levels are distinguished from one another by the degree of complexity in the polypeptide chain.

A single protein molecule may contain one or more of these protein structure types. The structure of a protein determines its function. For example, collagen has a super-coiled helical shape. It is long, stringy, strong, and resembles a rope, which is great for providing support. Hemoglobin, on the other hand, is a globular protein that is folded and compact. Its spherical shape is useful for maneuvering through blood vessels.

In some cases, a protein may contain a non-peptide group. These are called cofactors and some, such as coenzymes, are organic. Others are an inorganic group, such as a metal ion or iron-sulfur cluster.

Protein Synthesis

Proteins are synthesized in the body through a process called translation. Translation occurs in the cytoplasm and involves the translation of genetic codes into proteins.

The gene codes are assembled during DNA transcription, where DNA is transcribed into an RNA transcript. Cell structures called ribosomes help translate the gene codes in RNA into polypeptide chains that undergo several modifications before becoming fully functioning proteins.

Extracellular Matrix

The extracellular matrix is a meshwork of proteins and carbohydrates that binds cells together or divides one tissue from another. The extracellular matrix is the product principally of connective tissue, one of the four fundamental tissue types, but may also be produced by other cell types, including those in epithelial tissues. In the connective tissue, matrix is secreted by connective tissue cells into the space surrounding them, where it serves to bind cells together. The extracellular matrix forms the basal lamina, a complex sheet of extracellular matrix molecules that separates different tissue types, such as binding the epithelial tissue of the outer layer of skin to the underlying dermis, which is connective tissue. Cartilageis a connective tissue type that is principally composed of matrix, with relatively few cells.

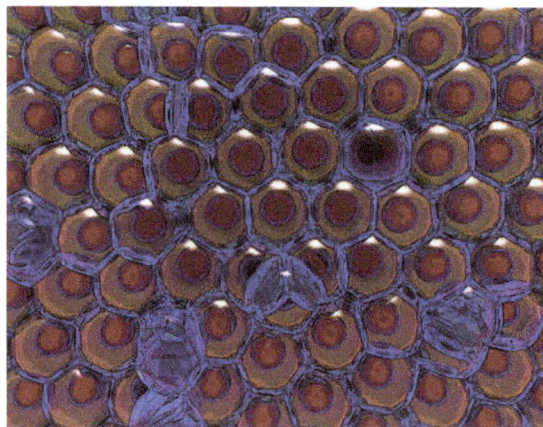

Collagens

Collagens are the principal proteins of the extracellular matrix. They are structural proteins that provide tissues with strength and flexibility, and serve other essential roles as well. They are the most abundant proteins found in many vertebrates. There are at least nineteen collagen family members whose subunits, termed α chains, are encoded by at least twenty-five genes. The primary protein sequence of all collagen subunits contains repeating sequences of three amino acids , the first being glycine with the second and third being any amino acid residue (sometimes referred to as a GLY–X–Y motif).

Most, if not all, collagens assemble as trimmers, with three subunits coming together to form a tightly coiled helix that confers rigidity on each collagen molecule. Assembly of the collagen trimer occurs in the cell by a self-assembly process, which is mediated by short amino acid sequences at both ends of each subunit, called propeptides. Some collagens, most notably collagen types I, II, III, and V, assemble into large, ropelike macrofibrils once they are secreted into the extracellular matrix. In these cases, the propeptides are cleaved off following secretion, permitting the trimeric molecules to undergo further self-assembly into fibrils. In the electron microscope each of these macrofibrils has a characteristic banded appearance and can be very large (up to 300 nanometers in diameter).

Type IV collagen, which is found in the basal lamina, does not assemble into a fibril since its subunits retain their propeptides following secretion from a cell. Its triple helix has a series of interruptions in the GLY−X−Y repeating motif, preventing the subunits from binding quite as tightly, and giving the molecule more flexibility. Type IV collagen forms a scaffold around which other basal lamina molecules assemble. In contrast to the fibril-forming collagens and type IV collagen, type XVII collagen is membrane-spanning protein. It is a component of a cell/matrix junction called the hemidesmosome.

The fibrillar collagens are also associated with a class of collagen molecules that themselves do not form fibrils but that appear to play an important role in organizing the highly ordered arrays of collagen fibrils that occur in some connective tissues. Examples of this collagen class include type IX and type XII collagen.

Collagens do not simply provide filler for tissues. Both fibrillar and basal lamina collagens interact with other extracellular matrix proteins and play important roles in regulating the activities of the cells with which they interact. Cells associate with collagen via cell surface receptors, and through such interactions collagens may have a profound impact on cell proliferation, migration, and differentiation. Fibers and meshworks of collagen molecules also act as a repository of growth factors and matrix-degrading enzymes. These are often present in inactive form and become activated in order for tissues to undergo remodeling, for example in development, during cyclical changes in the female reproductive system, and in pathological conditions such as cancer.

References

- Nucleic-acids, wmopen-biology: lumenlearning.com, Retrieved 25 August, 2019
- Carbon, apbiolecture, mancuso, science: guam.net, Retrieved 31 July, 2019
- Bio-aminoacid: chem4kids.com, Retrieved 17 February, 2019
- Protein-function: thoughtco.com, Retrieved 31 July, 2019
- Extracellular-matrix: biologyreference.com, Retrieved 25 August, 2019

Chapter 3

Cellular Anatomy

The anatomy of the cell is made up of a number of components such as cell membrane, cell organelles, cell wall, cytoplasm, cytoskeleton and cytosol. The topics elaborated in this chapter will help in gaining a better perspective about the anatomy of the cell as well as these components.

Cell Structure and Function

Cell Theory

The cell theory developed in 1839 by microbiologists Schleiden and Schwann describes the properties of cells. It is an explanation of the relationship between cells and living things. The theory states that:

- All living things are made of cells and their products.

- New cells are created by old cells dividing into two.

- Cells are the basic building blocks of life.

The cell theory applies to all living things, however big or small. The modern understanding of cell theory extends the concepts of the original cell theory to include the following:

- The activity of an organism depends on the total activity of independent cells.

- Energy flow occurs in cells through the breakdown of carbohydrates by respiration.

- Cells contain the information necessary for the creation of new cells. This information is known as 'hereditary information' and is contained within DNA.

- The contents of cells from similar species are basically the same.

Cells are the smallest form of life; the functional and structural units of all living things. Your body contains several billion cells, organised into over 200 major types, with hundreds of cell-specific functions.

Some functions performed by cells are so vital to the existence of life that all cells perform them (e.g. cellular respiration). Others are highly specialised (e.g. photosynthesis).

In figure shows a two-dimensional drawing of an animal cell. The diagram shows the structures visible within a cell at high magnification. The structures form the ultrastructure of the cell.

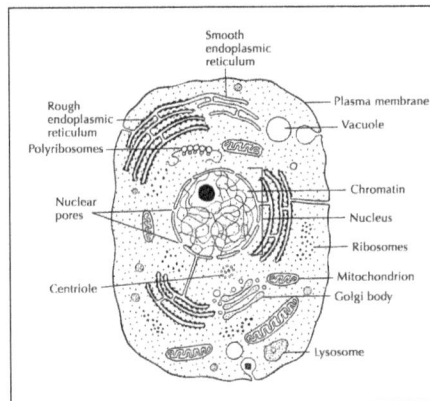

Diagram of the cell ultrastructure of an animal cell

We will now look at some of the basic cell structures and organelles in animal and plant cells.

Cell Wall

The cell wall is a rigid non-living layer that is found outside the cell membrane and surrounds the cell. Plants, bacteria and fungi all have cell walls. In plants, the wall is comprised of cellulose. It consists of three layers that help support the plant. These layers include the middle lamella, the primary cell wall and the secondary cell wall.

- Middle lamella: Separates one cell from another. It is a thin membranous layer on the outside of the cell and is made of a sticky substance called pectin.

- Primary cell wall: Is on the inside of the middle lamella and is mainly composed of cellulose.

- Secondary cell wall: Lies alongside the cell membrane. It is made up of a thick and tough layer of cellulose which is held together by a hard, waterproof substance called lignin. It is only found in cells which provide mechanical support in plants.

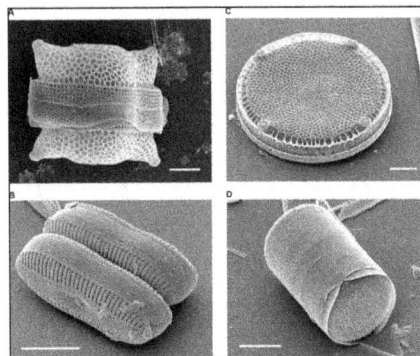

Scanning electron microscope micrographs of diatoms showing the external appearances of the cell wall. Scale bar: A, B, D: 10 um, C 20 um

Functions of the Cell Wall

- The main function of the wall is to protect the inner parts of the plant cell, it gives plant cells a more uniform and regular shape and provides support for the plant body.

- The cell wall is completely permeable to water and mineral salts which allows distribution of nutrients throughout the plant.

- The openings in the cell wall are called plasmodesmata which contain strands of cytoplasm that connect adjacent cells. This allows cells to interact with one another, allowing molecules to travel between plant cells.

Cell Membrane

The cell membrane, also called the plasma membrane, physically separates the intracellular space (inside the cell) from the extracellular environment (outside the cell). All plant and animal cells have cell membranes. The cell membrane surrounds and protects the cytoplasm. Cytoplasm is part of the protoplasm and is the living component of the cell.

The cell membrane is composed of a double layer (bilayer) of special lipids (fats) called phospholipids. Phospholipids consist of a hydrophilic (water-loving) head and a hydrophobic (water-fearing) tail. The hydrophobic head of the phospholipid is polar (charged) and can therefore dissolve in water. The hydrophobic tail is non-polar (uncharged), and cannot dissolve in water.

The lipid bilayer forms spontaneously due to the properties of the phospholipid molecules. In an aqueous environment, the polar heads try to form hydrogen bonds with the water, while the non-polar tails try to escape from the water. The problem is solved by the formation of a bilayer because the hydrophilic heads can point outwards and from hydrogen bonds with water, and the hydrophobic tails point towards one another and are 'protected' from the water molecules.

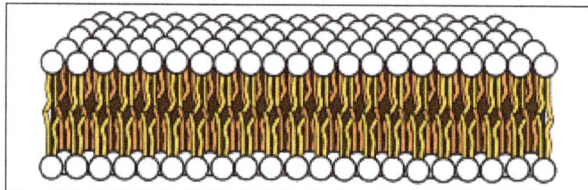

The lipid bilayer showing the arrangement of phospholipids, containing hydrophilic, polar heads and hydrophobic, non-polar tails

All the exchanges between the cell and its environment have to pass through the cell membrane. The cell membrane is selectively permeable to ions (e.g. hydrogen, sodium), small molecules (oxygen, carbon dioxide) and larger molecules (glucose and amino acids) and controls the movement of substances in and out of the cells. The cell membrane performs many important functions within the cell such as osmosis, diffusion, transport of nutrients into the cell, processes of ingestion and secretion. The cell membrane is strong enough to provide the cell with mechanical support and flexible enough to allow cells to grow and move.

Structure of the Cell Membrane: The Fluid Mosaic Model

S.J. Singer and G.L. Nicolson proposed the Fluid Mosaic Model of the cell membrane in 1972. This model describes the structure of the cell membrane as a fluid structure with various protein and carbohydrate components diffusing freely across the membrane. The structure and function of each component of the membrane is provided in the table below. Table refers to the components of the cell membrane shown in the diagram in figure.

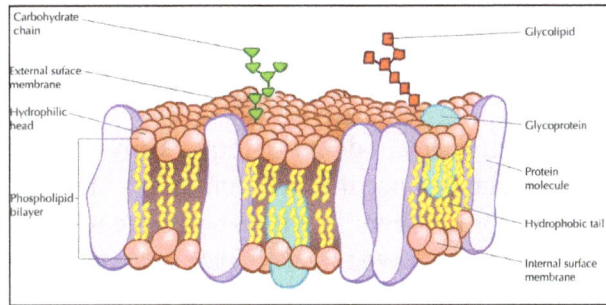

Fluid mosaic model of the cell membrane

Table: Structure and function of components of the cell membrane.

Component	Structure	Function
Phospholipid bilayer	Consists of two layers of phospholipids. Each phospholipid has a polar, hydrophilic (water-soluble) head as well as a non-polar, hydrophobic (water-insoluble) tail.	It is a semi-permeable structure that does not allow materials to pass through the membrane freely, thus protecting the intra and extracellular environments of the cell.
Membrane proteins	These are proteins found spanning the membrane from the inside of the cell (in the cytoplasm) to the outside of the cell. Membrane proteins have hydrophilic and hydrophobic regions that allow them to fit into the cell membrane.	Act as carrier proteins which control the movement of specific ions and molecules across the cell membrane.
Glycoproteins	Consist of short carbohydrate chains attached to polypeptide chains and are found on the extracellular regions of the membrane.	These proteins are useful for cell-to-cell recognition.
Glycolipids	Carbohydrate chains attached to phospholipids on the outside surface of the membrane.	Act as recognition sites for specific chemicals and are important in cell-to-cell attachment to form tissues.

Movement across Membranes

Movement of substances across cell membranes is necessary as it allows cells to acquire oxygen and nutrients, excrete waste products and control the concentration of required substances in the cell (e.g. oxygen, water, hormones, ions, etc.). The key processes through which such movement occurs include diffusion, osmosis, facilitated diffusion and active transport.

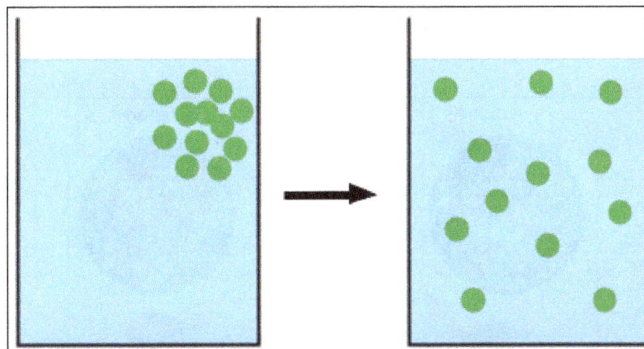

Diffusion is the movement of molecules from a region of higher concentration to a lower concentration. It is a passive process (i.e. does not require input of energy).

Diffusion is the movement of substances from a region of high concentration to low concentration. It is therefore said to occur down a concentration gradient. The diagram below shows the movement of dissolved particles within a liquid until eventually becoming randomly distributed.

Diffusion is a passive process which means it does not require any energy input. It can occur across a living or non-living membrane and can occur in a liquid or gas medium. Due to the fact that diffusion occurs across a concentration gradient it can result in the movement of substances into or out of the cell. Examples of substances moved by diffusion include carbon dioxide, oxygen, water and other small molecules that are able to dissolve within the lipid bilayer.

Osmosis

When the concentration of solutes in solution is low, the water concentration is high, and we say there is a high water potential. Osmosis is the movement of water from a region of higher water potential to a region of lower water potential across a semi-permeable membrane that separates the two regions. Movement of water always occurs down a concentration gradient, i.e. from higher water potential (dilute solution) to lower potential (concentrated solution). Osmosis is a passive process and does not require any input of energy. Cell membranes allow molecules of water to pass through, but they do not allow molecules of most dissolved substances, e.g. salt and sugar, to pass through. As water enters the cell via osmosis, it creates a pressure known as osmotic pressure.

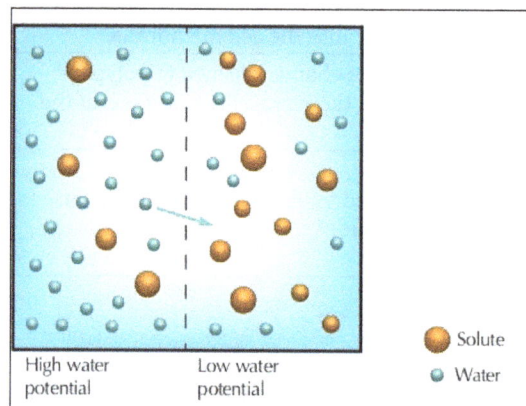

Osmosis is the movement of water from an area of high water potential to an area of low water potential across a semi-permeable membrane.

In biological systems, osmosis is vital to plant and animal cell survival. Figure demonstrates how osmosis affects red blood cells when they are placed in three different solutions with different concentrations.

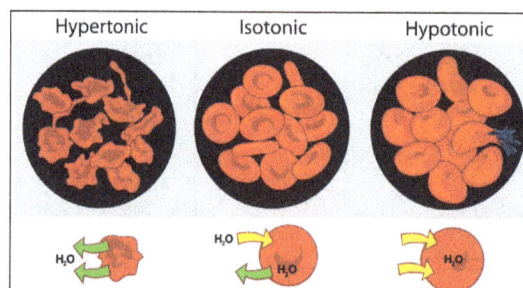

The effect of hypertonic, isotonic and hypotonic solutions on red blood cells.

Hypertonic (concentrated)	Isotonic	Hypotonic (dilute)
The medium is concentrated with a lower water potential than inside the cell, therefore the cell will lose water by osmosis.	The water concentration inside and outside the cell is equal and there will be no net water movement across the cell membrane. (Water will continue to move across the membrane, but water will enter and leave the cell at the same rate.)	The medium has higher water potential (more dilute) than the cell and water will move into the cell via osmosis, and could eventuality cause the cell to burst.

Plant cells use osmosis to absorb water from the soil and transport it to the leaves. Osmosis in the kidneys keeps the water and salt levels in the body and blood at the correct levels.

Facilitated Diffusion

Facilitated diffusion is a special form of diffusion which allows rapid exchange of specific substances. Particles are taken up by carrier proteins which change their shape as a result. The change in shape causes the particles to be released on the other side of the membrane. Facilitated diffusion can only occur across living, biological membranes which contain the carrier proteins. A substance is transported via a carrier protein from a region of high concentration to a region of low concentration until it is randomly distributed. Therefore movement is down a concentration gradient.

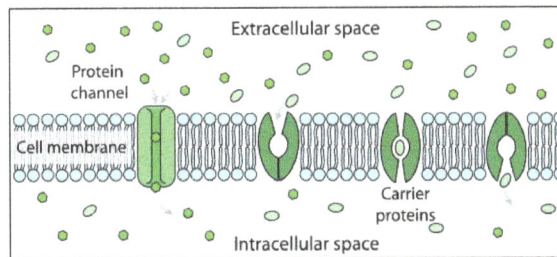

Facilitated diffusion in cell membrane, showing ion channels and carrier proteins.

Examples of substances moved via facilitated diffusion include all polar molecules such as glucose or amino acids.

Active Transport

Active transport is the movement of substances *against* a concentration gradient, from a region of *low concentration* to *high concentration* using an input of energy. In biological systems, the form in which this energy occurs is adenosine triphosphate (ATP). The process transports substances through a membrane protein. The movement of substances is selective via the carrier proteins and can occur into or out of the cell.

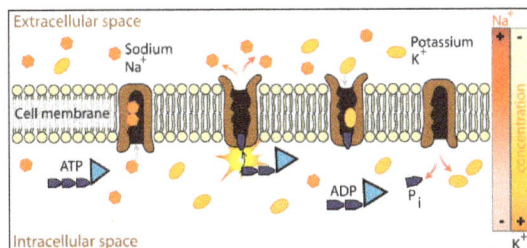

The sodium-potassium pump is an example of primary active transport.

Examples of substances moved include sodium and potassium ions as shown in figure.

Cell Membrane

The cell membrane (plasma membrane) is a thin semi-permeable membrane that surrounds the cytoplasm of a cell. Its function is to protect the integrity of the interior of the cell by allowing certain substances into the cell while keeping other substances out. It also serves as a base of attachment for the cytoskeleton in some organisms and the cell wall in others. Thus the cell membrane also serves to help support the cell and help maintain its shape.

Molecules that can readily cross cell membranes are frequently needed in biological research and medicine. Permeable molecules that are useful for biological research include indicators of ion concentrations and pH, fluorescent dyes, crosslinking molecules, fluorogenic enzyme substrates, and various protein inhibitors. In medicine, numerous drugs are small molecules acting on intracellular targets, such as statins that inhibit cholesterol production, and reverse transcriptase inhibitors used for the treatment of HIV. Given the high level of interest across multiple areas of study in modulating intracellular targets, a broad overview of cytosolic delivery strategies could contribute to orienting researchers newly entering the field, and bringing together the solutions that have been proposed for various cargos.

Cellular Organization

Before discussing membrane transport mechanisms, we will briefly describe the structure of the mammalian plasma membrane and components of the endocytic and secretory pathways. For the purpose of intracellular delivery, it is important to note that the interior of endocytic vesicles and the lumen of the organelles involved in the secretory pathway topologically correspond to the extracellular space.

Structure and Organization of the Cell Membrane

The plasma membrane is a complex composite of multiple lipid species and membrane proteins. Three major classes of lipids, including glycerophospholipids, sphingolipids and cholesterol, form a bilayer approximately 5nm in width. Spatially, these lipids are distributed asymmetrically across the bilayer. Additionally, according to the lipid raft hypothesis, the membrane is thought to contain lateral organizations enriched in sphingolipids, cholesterol, and glycosylphosphatidylinositol (GPI)-anchored proteins.

The ratio of protein to lipid in cellular membranes has been approximated to be 1:40 by number, suggesting that the membrane may in fact be crowded with proteins. This ratio can vary substantially by cell type, where metabolically active membranes are richer in protein. Membrane proteins can actively influence the organization of the membrane by forming specific and nonspecific interactions with lipids in the immediate boundary.

Finally, the plasma membrane is in continuous motion, creating a highly dynamic structure. In addition to lateral diffusion, phospholipid flip-flop for some lipids is thought to occur on the order of minutes, and faster so for cholesterol. Also, cells constantly internalize and recycle their membranes.

Secretory Pathway

In mammalian cells, secretory proteins are translocated across the Endoplasmic Reticulum (ER) membrane into the lumen co-translationally via the translocon complex. Misfolded proteins in the ER are transported back to the cytosol and degraded by the proteasome, a process termed ER-associated degradation (ERAD). Correctly folded proteins are transported across the Golgi network and released into the extracellular space via secretory vesicles. Specialized vesicles also mediate retrograde transport from the golgi to the ER, and from older to newer golgi.

Endocytic Pathways

Multiple endocytic pathways facilitate the internalization of exogenous cargo, creating a complex web of intracellular traffic. The choice of which endocytic pathway is utilized may depend on the cargo. Nonspecific internalization of large volumes of fluid-pinocytosis-occurs in all cells, typically triggered by external stimuli such as growth factors. Clathrin-dependent and independent routes of endocytosis generate primary endocytic vesicles that subsequently fuse with early endosomes, a major sorting station. Traveling down tracks of microtubules towards the perinuclear space, the early endosomes mature into multivesicular bodies (MVB), late endosomes and lysosomes. Endocytosed material that has not been recycled to the plasma membrane or exchanged with the *trans* golgi network is proteolyzed by hydrolytic enzymes in the lysosome.

Methods to Measure Membrane Permeation

The permeability of a given molecule can be quantitatively represented by its permeability coefficient (typically in units of cm/s), which is a measure of how fast it can cross a membrane. High-throughput methods to measure the permeability coefficients of small molecules are routinely performed in drug development. Artificial lipid bilayers of various compositions or cell monolayers (typically the colorectal Caco-2 or renal MDCK cell lines) are widely used as model barriers. While the former allows passive permeation only, the latter also allows transporter-mediated permeation. To disentangle these two modes of transport, cell lines that lack certain transporters, such as P-glycoprotein, have been developed.

Table: Permeability coefficients of select molecules.

Species	Molecule	Permeability coefficient (cm/s)	Membrane type
Ions	Na^+ K^+	5.0×10^{-14} 4.7×10^{-14}	Artificial membrane
Small molecules	O_2	2.3×10^1	Artificial membrane
	CO_2	3.5×10^{-1}	Artificial membrane
	H_2O	3.4×10^{-3}	Artificial membrane
	EtOH	2.1×10^{-3}	Erythrocyte membrane
	Steroids	10^{-3} to 10^{-4}	Cell monolayer
	Urea	4.0×10^{-6}	Artificial membrane
	Glycerol	5.4×10^{-6}	Artificial membrane
	Small molecule drugs	10^{-5} to 10^{-6}	Artificial membrane
Peptides	Cyclosporin A	2.5×10^{-7}	Artificial membrane
	TAT	2.7×10^{-9}	Artificial membrane

In contrast to small molecules, the permeability of peptides or proteins across model membranes is rarely reported, reflecting in part the difficulty of translocating such molecules across a lipid membrane. It is technically challenging to accurately quantify the number of functional peptides or proteins that have successfully entered the cytoplasm. Selective isolation of the cytosol (and not endosomal compartments) using cellular fractionation or digitonin-mediated permeabilization of the plasma membrane have been reported. Immunoprecipitation demonstrating the intended disruption of intracellular protein–protein interactions has also been presented as evidence of permeation.

Fluorescence microscopy-based methods or biological assays measuring the activity of the payload in the cytoplasm are also employed. In microscopy, diffuse cytosolic staining (indicating endosomal release) is contrasted with punctate signal (indicating endosomal entrapment) to provide a qualitative assessment of permeation. However, it should be noted that payloads in the cytoplasm may also aggregate or associate with subcellular organelles to produce punctate patterns. In some cases, automated image analyses have been reported to identify endosomal release events. In any case, the presence of labeled payload in the cytoplasm does not guarantee that it has retained its function, and the label itself or fixation steps may cause artifacts in cellular distribution. All observations are subject to the detection threshold of the instrument. Flow cytometry may be used as an alternative to microscopy if the fluorescence spectra are distinct in the endosomal compartment and the cytoplasm.

Alternatively, cytosolic uptake can be confirmed by measuring a biological effect that is generated only when the payload is in the cytoplasm. For example, peptides have been conjugated to dexamethasone (Dex) derivatives, which bind to transiently expressed gluocococorticoid receptor (GR)-fusion proteins in the cytosol to induce a reporter or alter its localization. It should be noted that reporter gene expression inherently amplifies the signal through multiple rounds of transcription and translation. In cases where the biological activity of the pay-load is reported, certain payloads can generate the measured macroscopic effect with fewer numbers. This is particularly true for catalytic proteins. For example, approximately 50 molecules of β-lactamase in a single cell have been reported to generate a detectable signal from catalyzing a fluorogenic substrate, albeit over a long period of time (16 h). Similarly, in theory, four molecules of Cre recombinase can repeatedly catalyze multiple recombination events to promote recombined gene expression. Single molecules of toxins such as diphtheria and ricin have been estimated to kill a cell.

Natural Membrane Transport Mechanisms

Small, moderately polar molecules are able to passively diffuse across the cell membrane. To transport larger, more polar compounds such as most sugars, amino acids, peptides, and nucleosides, membrane transporters are utilized. Interestingly, bacteria and viruses have developed sophisticated mechanisms to transport whole organisms, protein toxins, or genetic material into the mammalian cytoplasm.

Passive Diffusion

Passive diffusion across a cellular membrane is driven by the concentration and electric gradient of the solute and does not require the use of energy. In the simplest terms, passive diffusion is considered a three-step process, where the permeant first partitions into the membrane, diffuses across, and is released into the cytosol (known as the homogeneous solubility-diffusion model).

The most important parameters that govern transmembrane diffusion are polarity and size. For example, small nonpolar gases such as O_2, CO_2, and N_2, and small polar molecules such as ethanol cross lipid membranes rapidly. High permeability coefficients have been reported for such molecules across artificial lipid membranes, such as 2.3×10^1 cm/s for O_2 and 3.5×10^{-1} cm/s for CO_2. The small, but highly polar water molecule is still able to diffuse across artificial membranes rapidly with a permeability coefficient of 3.4×10^{-3} cm/s.

In comparison, even slightly larger polar metabolites such as urea and glycerol have lower permeability across artificial membranes (approximately 10^{-6} cm/s). The plasma membrane is virtually impermeable against larger, uncharged polar molecules and all charged molecules including ions. Indeed, despite their small size, Na^+ and K^+ have extremely low permeability coefficients (approximately 10^{-14} cm/s). Apart from small solutes of moderate polarity, the number of natural molecules known to passively diffuse across the cell membrane is surprisingly limited. Steroid hormones have been assumed to do so, although direct experimental evidence is scarce. Permeability coefficients on the order of 10^{-4} cm/s have been reported for a number of steroids across cell monolayers.

Interestingly, some non-endogenous natural products have been proposed to passively diffuse across the cell membrane despite their relatively higher polarity and size, such as the cyclic peptide Cyclosporin A (CsA). Prescribed as an immunosuppressant, its intracellular mode of action and low EC$_{50}$ in cells (7–10 nM) suggests that CsA is capable of passively permeating the cell membrane. Still, the reported permeability coefficient of CsA—2.5×10^{-7} cm/s across artificial membranes -is relatively low compared to those of small molecules that are considered highly permeable (on the order of 10^{-5} cm/s or higher).

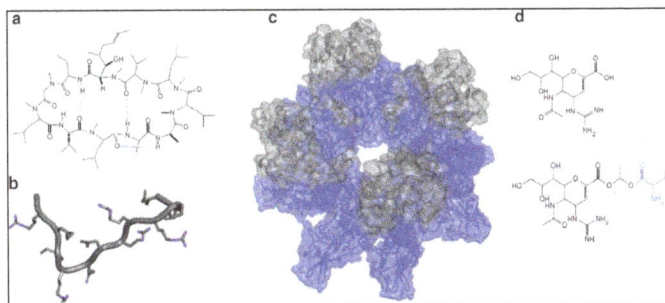

In figure, Cyclosporin A (CsA) in its closed conformation in nonpolar solvent. The four intramolecular hydrogen bonds (*dotted lines* in *blue*) are thought to shield the polarity of the molecule. (b) The TAT peptide segment excerpted from the NMR structure of HIV-1 TAT protein (adapted from PDB 1TIV). The guanidinium nitrogens (*blue*) are thought to enhance the interaction between TAT and the cell membrane. (c) A slanted top- down view of the pre-pore formed by anthrax toxin protective antigens (PAs) (*blue*) in complex with lethal factors (LFs) (*gray*), which are translocated across the full pore. Shown in the figure are eight molecules of PA bound to four molecules of LF (PA$_8$(LF$_N$)$_4$) (PDB 3KWV) . (d) The neuraminidase inhibitor Zanamivir (*top*) and Zanamivir-L-Val (*bottom*). The conjugated valine (*blue*) has been proposed to render Zanamivir into a substrate for amino acid transporters.

Transporter-mediated Entry

To facilitate the entry or export of molecules that are insufficiently permeable, cells utilize membrane transporters, the expression of which may depend on cell type. Active transporters use energy to translocate substrates against their concentration gradients, whereas passive transporters allow transmembrane diffusion without additional energy. Approximately 10 % of all human genes are transporter related, emphasizing their functional significance. In the following, a selection of transporters is described, ordered according to the size of the substrate.

Ion channels allow the passive diffusion of inorganic ions with high specificity, often in response to stimuli such as changes in transmembrane potential, ligands, light, or mechanical stress. Alternatively, ions can also be actively transported by ion pumps, such as the sodium/potassium pump (the Na^+, K^+-ATPase), which transports three Na^+ ions extracellularly and two K^+ ions intracellularly for every molecule of ATP hydrolyzed. Microbe-synthesized ionophores, such as valinomycin, facilitate the diffusion of ions across the cell membrane by complexing and shuttling ions. Other ionophores such as gramicidin A form channels.

Small molecules are also often transported. Water is transported across the membrane by the aquaporin (AQP) family of passive channels. Aquaporins have been reported to transport other gases and solutes as well, such as CO_2, NO, H_2O_2, arsenite, ammonia (in addition to the Rh proteins, urea (in addition to the urea transporters) and glycerol. (This is an abbreviated list excerpted from Bienert et al.).

Sugars, including glucose, galactose, and fructose, are molecules of high polarity and intermediate size, and are imported into the cell by the glucose transporter (GLUT) family of facilitated transporters. Most amino acids are transported across the cell membrane by secondary active transporters that utilize the energy stored in the electrochemical gradient of another solute Nucleobases and nucleosides also have associated secondary transporters on the plasma membrane. Di- and tri-peptides of various chemical character are transported by the oligopeptide transporter PepT1, which has been reported to transport neutral tripeptide-like β-lactam antibiotics and peptide-like drugs as well. Alternatively, α-Amanitin, a cyclic octapeptide that inhibits eukaryotic RNA polymerase II, has been reported to enter cells via an organic anion transporting polypeptide (OATP) transporter.

To note, transporters may mediate the efflux of molecules as well. A variety of structurally unrelated compounds, including small-molecule drugs, are substrates of efflux pumps in the ATP-binding cassette (ABC) transporter family such as the multidrug resistance protein (MRP) family , the P-glycoprotein pump (P-gp, also known as multidrug resistance protein 1(MDR1)) , and the breast cancer resistance protein (BCRP).

Other Methods of Cytosolic Entry

A majority of the examples discussed in the following first involve the cargo being internalized into the cell via various endocytic pathways. Reiterating an earlier point, endocytosed cargo are topologically still in an extracellular space separated from the cytoplasm by a lipid membrane. Thus, an additional "endosomal escape" (or "endosomal release") step is required where the cargo is transported across the membrane to access the cytoplasm. Some peptidic, viral, or bacterial components are thought to accomplish this step, not through passive diffusion or active transport, but by disrupting cellular membranes, allowing the passage of large and charged compounds. The mechanisms of most such processes are not yet fully elucidated and subjects of active research.

Peptides

Cell-penetrating peptides (CPP), also known as peptide transduction domains (PTD), are a diverse class of peptides that have been reported to traverse the cell membrane. Representative members of this family such as the Trans-Activator of Transcription (TAT) peptide and penetratin were initially identified as segments within naturally occurring proteins with proposed membrane permeability, such as homeoproteins.

The mechanisms of how these peptides cross the cell membrane is still unclear and likely differs amongst peptides. The TAT peptide for example (GRKKRRQRRRPSQ) is rich in arginines, and the delocalized positive charge in their guanidinium moieties has been proposed to allow extensive interactions with negatively charged cell membranes. TAT is thought to bind to the glycosaminoglycans (GAG) on the cell surface such as heparin sulfate or adsorb into the glycerol backbone region of the lipid bilayer, eventually being macropinocytosed. Various models of TAT-mediated perturbations of the cell membrane have been proposed, including the formation of transient pores.

The reported permeability coefficient of TAT across artificial membranes is very low at 2.7×10^{-9} cm/s, which may reflect its need for structural features specific to the cell membrane to be able to translocate. Typically, relatively high (μM) concentrations of TAT are required in vitro to observe translocation, and the efficiency of such may depend on the cell line. Also, as mentioned earlier, fixation of cells treated with fluorescently labeled TAT may lead to artifacts in cellular distribution, and thus reported results need to be interpreted with caution.

Protein Toxins

A number of plant and bacterial toxins are potent inhibitors of central cellular functions such as protein synthesis. However, before they can have their effect, they must gain access to their cytosolic targets. Typically, a separate domain (typically denoted the B domain, translocation domain, or translocation complex (when an oligomer)) is responsible for binding to cellular receptors and translocating the catalytic domain (the A domain) into the cytoplasm.

Some toxins form their own pores, such as the diphtheria and anthrax toxins. The translocation domains of anthrax toxin, known as protective antigen (PA), oligomerizes into a prepore complex following proteolytic activation. Subsequent internalization and endosomal acidification is thought to trigger its conversion into a full pore, through which catalytic domains escape into the cytosol.

A number of other toxins, such as the plant toxin ricin, *Pseudomonas* exotoxin A, and cholera toxin, take advantage of the ERAD machinery to enter the cell. Following binding to gangliosides via its B domains, cholera toxin is internalized and trafficked to the ER where the A domain is reduced and unfolded. This domain is subsequently refolded in the cytoplasm.

Viruses

Some viruses enter the cell through the plasma membrane, but more commonly from endocytic compartments after binding to cellular receptors and triggering various endocytic pathways. Viruses can be classified into enveloped viruses, which are encased in a lipid membrane containing glycoproteins, or non- enveloped viruses, which lack a membrane.

In general, enveloped viruses are thought to orchestrate the fusion of host and viral membranes using viral fusion proteins, which expose hydrophobic peptides upon environmental triggers such as receptor binding, low pH, or proteolytic cleavage. For example, influenza A exposes a hydrophobic segment of hemagglutinin (HA) upon endosomal acidification. With this mechanism, there is no need to translocate across the cell membrane.

Non-enveloped viruses, in contrast, have to cross the membrane in order to access the cytoplasm. In general, these viruses are thought to mediate the disruption of cellular membranes by exposing or releasing lytic peptides that are amphipathic or hydrophobic. Alternatively, members of the polyomavirus family such as the simian virus (SV40) use a strategy similar to the aforementioned cholera toxin, and hijack the ERAD machinery.

Approaches to Design and Improve Membrane Permeability

Small Molecule Cargo

Decades of pharmaceutical research have provided design principles that maximize the chance of obtaining a drug able to efficiently distribute within an organism and permeate through cell membranes. While bioavailability is often the reported parameter of interest, efficient membrane permeation is likely necessary for bio-availability. Therefore, rules that have been devised in medicinal chemistry to achieve favorable bioavailability are a reasonable guide for the design of membrane-permeating small molecules.

Predicting Passive Permeation

Lipinski's "Rule of 5" has been the most influential framework correlating the physicochemical properties of a given compound with its membrane permeability and bioavailability in the context of small-molecule drug development . It postulates that poor absorption or permeation is more likely when:

(1) The calculated lipophilicity (clogP) is over 5;

(2) The molecular weight is over 500;

(3) There are more than five hydrogen bond donors (well represented by the sum of OH and NH bonds); and

(4) There are more than ten hydrogen bond acceptors (represented roughly, by the sum of Ns and Os).

The Rule of 5 has been generally successful at predicting membrane permeability, but not all compounds that comply with the rules are permeable, and permeable compounds that deviate from the rules are not uncommon. Nonetheless, as suggested by Guimarães et al., the Rule of 5 does identify key physicochemical parameters, namely the polarity, size, and lipophilicity of the permeant, that are important for passive diffusion. These interrelated factors can affect the partitioning, diffusion, or both, of the molecule into and across the membrane.

Alternative metrics for these parameters have also been proposed. Regarding polarity, the polar surface area (PSA) of a compound has been used in addition to the number of hydrogen bond donors and acceptors. For molecular size, studies have inversely-correlated the permeability of small solutes with molecular volume or cross- sectional area. A different but related parameter, the number of rotatable bonds, has been suggested as well, where molecules with fewer rotatable bonds and lower PSA were reported to have better permeability across artificial membranes. Additionally, it has also been proposed that conformationally flexible molecules that are able to form intramolecular hydrogen bonds in a low dielectric environment may adaptively reduce their surface polarity for improved permeation. Unsurprisingly, even if the hydrogen bond counts or PSA is low, localized charge or highly polar groups can significantly decrease the permeability of an otherwise permeable parent compound by orders of magnitude.

Beyond empirical correlations, molecular dynamics (MD) simulations are increasingly applied to calculate the energetic barrier of transmembrane diffusion, from which permeability coefficients can be derived. Improved computational power and coarse-grained modeling have reduced computing time. However, although these methods are invaluable in estimating permeabilities that are difficult to obtain experimentally, utilizing them on a routine basis is yet hampered by the computational cost and the effort involved in building a suitable representation of the molecule of interest. Estimates for large molecules may be particularly prone to inaccuracy due to insufficient sampling of their conformational space during the simulations.

Predicting Transporter-mediated Permeation

Designing compounds to be substrates of a specific transporter is currently difficult, although indirect approaches have been proposed to identify metabolites that are structurally similar to a given compound. Alternatively, conjugating compounds to known transporter substrates such as amino acids has been reported to improve permeation and oral adsorption by engaging PepT. In such "prodrug" approaches, the conjugated substrates are designed to be cleaved intracellularly or during circulation to release the free drug. Although designing specific transporter substrates is infeasible at the moment, it should be kept in mind that transporters can affect a compound's permeation.

Comparing Theory and Empirical Data for Molecular Probes

Empirical permeability data from molecular probes and labeling molecules roughly agree with the theoretical expectations. Generally, small and uncharged fluorophores, and those whose charge is delocalized over the fluorophore (e.g., TAMRA), are sufficiently membrane-permeable to be used in intra-cellular protein labeling applications. However, fluorescent dyes carrying localized charges (e.g., the sulfonic acid derivatives of Cy3 or Cy5) display low membrane permeability. Esterification of charged groups is one strategy to mask the effects of charge.

An example of the size-dependence of membrane translocation is provided by fluorescent dyes modified with long and hydrophobic lipid-like tails. For the voltage-sensitive dyes Di-4-ANEPPS and Di-8-ANEPPS (equipped with two octyl and butyl chains, respectively), a strong decrease in membrane flip-flop was observed across planar black lipid membranes for the long-chain variant. A similar result was obtained for the dyes DiI-C12 and DiI-C18. The counterintuitive result where increasing the overall hydrophobicity of the molecules strongly reduced the rate of flip-flop is likely due to the concomitant increase in molecular size. A similar result has been reported with anthroyl fatty acids in liposomes, where the rate of flip-flop was observed to be 200-fold faster for a C11-fatty acid compared to a C18-fatty acid.

These molecular probes may also be substrates of cellular transporters. For example, acetoxy-methyl ester (AM) derivatives of various fluorescent indicators were observed to be actively exported from cells by multidrug transporters. Of note, passive diffusion and active transport may occur concomitantly. Chidley et al. studied the intracellular access of various organic molecules used for protein labeling via the SNAP-tag system in yeast strains that were either wild-type or had three efflux transporters deleted. The study showed a strong decrease in uptake with increasing size and polarity of the labeling molecule, suggesting entry by passive diffusion. Additionally, it showed that labeling efficiency increased in the modified yeast strain, presumably due to reduced active export.

Peptide Cargo

It is unlikely that peptides will passively diffuse across the cell membrane, but altering their physical properties (such as conformational flexibility and polarity) has been proposed to improve their permeability. Despite interesting findings-a selection of which is discussed in the following-conflicting experimental results have been reported. A straightforward method for converting a non-permeable peptide into an efficiently permeating entity is thus not available so far.

Addressing Conformation and Polarity

Macrocyclic drugs-those with ring architecture of 12 or more atoms, including cyclic peptides—tend to be larger and more polar than most small-molecule drugs, falling outside the Rule of 5. Yet some are administered orally, suggesting that they may be membrane permeable. In the case of cyclosporin A, this is believed to occur by passive diffusion.

Following such examples, cyclizing a given peptide and methylating select amide bond nitrogens have been proposed to improve its membrane permeation and/or bioavailability. Such modifications, when made judiciously , are thought to facilitate the formation of intramolecular hydrogen bonds in response to the low dielectric environment of the membrane interior Passive permeability values ranging from 6.3×10^{-7} cm/s to approximately 7.7×10^{-6} cm/s (estimated from) have been reported for certain hydrophobic cyclic peptides.

Alternatively, cyclization and amidation may alter a compound's specificity towards membrane transporters. In a study of 54 cyclic alanine hexapeptides containing various degrees of N-methylation, Ovadia et al. reported that none of the tested peptides showed permeation across artificial membranes. However, some peptides were found to be highly permeable across Caco-2 cell monolayers (on the order of 10^{-5} cm/s), suggesting that transporters may be involved.

In some instances, cyclization by covalently linking internal residues has been proposed to increase permeability by changing the peptide's α-helical content. Such modifications include hydrocarbon "staples" linking the side chains of nonnatural amino acids inserted into the peptide, and "hydrogen bond surrogates" replacing a main chain hydrogen bond with a carbon–carbon or disulfide bond. Such modifications have lead to the development of peptide inhibitors against intracellular targets such as the ICN1/CSL complex (involved in the NOTCH signaling pathway), Ras and MDM2/MDMX.

However, introduction of a staple alone does not guarantee an improvement in permeability.

Extensive optimization may still be required for multiple factors such as the position, length, and stereochemistry of the staple, as well as the charge and amino acid sequence of the peptide.

Designing Cell-penetrating Peptides

Extensive effort in discovering novel membrane-permeable peptides has generated significant diversity in the physicochemical character of reported CPPs. Methods have been proposed to synthetically design permeable peptides or predict such segments from a given protein sequence.

Introducing arginine residues within α-helices has been proposed to improve permeability. In a study of the avian pancreatic polypeptide (aPP), a 36-residue peptide/miniature protein, and CP1, a 28-residue zinc finger, substituting five residues within the α-helix with arginine increased the permeability to that comparable with TAT. The authors estimated that approximately 1–5 % of the internalized peptides were being released into the cytosol.

Protein Cargo

Proteins cannot passively diffuse across the cell membrane due to their size and polarity. Thus, a delivery system or technique is always required, similar to nucleic acid transfection. However, while nucleic acid transfection reagents are now routinely used in the laboratory, there are no equivalent standards for the delivery of proteins. In the following, we survey strategies that have been proposed to deliver proteins across the cell membrane. Given the physicochemical diversity of proteins and their delicate nature, it is challenging to design a system or method that is readily generalizable to multiple proteins while maintaining the cargo's respective function and stability.

Mechanical Disruption of the Membrane

Varying physical methods of disrupting the cell membrane, such as microinjection and electroporation, have been proposed for delivering compounds ranging from small molecules to proteins. Sharei et al. developed a microfluidic device that transiently disrupts the plasma membrane through physical constriction. Silicon "nanowires" that pierce the cell membrane have also been reported.

Peptide-based Strategies

CPPs have been reported to enhance the permeability of various macromolecules, including proteins. Early studies showed that the TAT peptide can mediate the translocation of covalently coupled proteins. In later studies, an amphiphilic CPP Pep-1 was reported to noncovalently complex and translocate peptide and protein cargos.

Substance P (SP), an 11-residue neuropeptide implicated in cancer progression, has been proposed to mediate the cytosolic delivery of synthetic antibody fragments and nucleic acids following covalent conjugation. Its natural GPCR partner, the neurokinin-1 receptor (NK1R), has been suggested to play a role in mediating uptake. The mechanisms by which such peptides mediate translocation remains to be clarified.

Protein-based Strategies

Various pore- or channel-forming proteins of bacterial origin have been utilized to translocate exogenous proteins. Highly sophisticated secretion systems, which transport proteins directly from the bacterial cytoplasm to the eukaryotic host's, have been reported to deliver proteins to the cytosol of antigen-presenting cells. Doerner et al. reported the functional expression of an engineered bacterial channel (MscL) in mammalian cells, the opening and closing of which could be controlled chemically. Alternatively, the cholesterol-dependent cytolysin (CDC) family of pore-forming toxins, which are capable of forming macropores up to 30 nm in diameter, have been proposed as "reversible permeabilization" reagents for delivering exogenous proteins.

In addition to pore- or channel-forming proteins, the membrane- translocating domains of bacterial toxins have been proposed as a modular tool that can be fused to, and enhance the intracellular delivery of, other proteins. In instances where the receptor-binding domain of the toxin is physically distinct from the translocation domain, the former has been replaced with alternative targeting moieties to generate immunotoxins. Immunotoxins retain the cytotoxicity of the parent toxin but are directed at specific cell types.

Additionally, "supercharged" GFP, a variant engineered to have high net positive charge (+36), and certain human proteins with naturally high positive charge have been reported to translocate across the cell membrane. Curiously, 3E10, an autoantibody proposed to bind to dsDNA, has been proposed to penetrate into the nucleus and impair DNA repair, or translocate an exogenous phosphatase across the cell membrane.

Virus-based Strategies

Packaging proteins in virus-like particles or attaching them to an engineered bacteriophage T4 head has been reported to enhance cytosolic delivery. In addition, although not yet utilized as a delivery system, it has been reported that virus-bound antibodies co-internalize into the cytoplasm along with the virus.

Lipid- and Polymer-based Strategies

With lipid-based materials, the protein cargo is either encapsulated in liposomes or complexed with lipids. Regarding the latter strategy, lipid formulations that have been successful in the transfection of DNA have been attempted for the protein delivery. For example, a formulation based on a mixture of cationic and neutral lipids was reported to translocate negatively charged proteins.

Similarly, polymer-based formulations that have been successfully used for nucleic acid transfections have also been examined for their ability to "transfect" proteins. The "proton sponge effect" is an influential hypothesis still undergoing debate, which states that materials such as polyethylenimine (PEI) that are rich in protonatable amines, will cause a significant buffering of protons and subsequent osmotic swelling in endosomes. The endosome is then proposed to stall its maturation and eventually rupture. Poly-β-amino esters (PBAEs), successfully developed for the transfection of nucleic acids, are thought to take advantage of this proton sponge effect. Su et al. reported that biodegradable PBAE nanoparticles enhance the endosomal escape of various cargos, including proteins, when co-administered.

Alternatively, Yan et al. reported a technology to encapsulate single proteins in a polymeric shell (termed "nanocapsule") after attaching the monomeric building blocks of the polymer directly to the protein. Such nanocapsules, designed to be degraded in response to environmental stimuli such as protease activity or changes in pH or redox potential, were reported to deliver proteins including transcription factors.

Inorganic Material-based Strategies

A variety of inorganic materials have also been proposed to translocate protein cargo, including silica, carbon nanotubes, quantum dots, and gold nanoparticles.

Cell Organelles

A small organ-like structure present inside the cell is called a cell organelle. It has a particular structural makeup and performs a specific function. Depending upon the presence or absence of membrane, cell organelles can be classified into three categories, namely:

- Without membrane: Some cell organelles like ribosomes are not bounded by any membrane. They are present in prokaryotic as well as eukaryotic organisms.

- Single membrane-bound: Some organelles are bounded by a single membrane. For example, vacuole, lysosome, Golgi apparatus, Endoplasmic Reticulum etc. They are present only in a eukaryotic cell.

- Double membrane-bound: Cell organelles like mitochondria and chloroplast are double membrane-bound organelles. They are present only in a eukaryotic cell.

Cell Nucleus

The cell nucleus is the site of many important biological functions of the eukaryotic cell. These processes include transcription, replication, splicing and ribosome biogenesis. The effect of these processes extends to affecting cellular metabolism and growth. The nucleus contains approximately 2m of DNA which is enmeshed by the nuclear envelope, a cross-linked network of proteins and membranes.

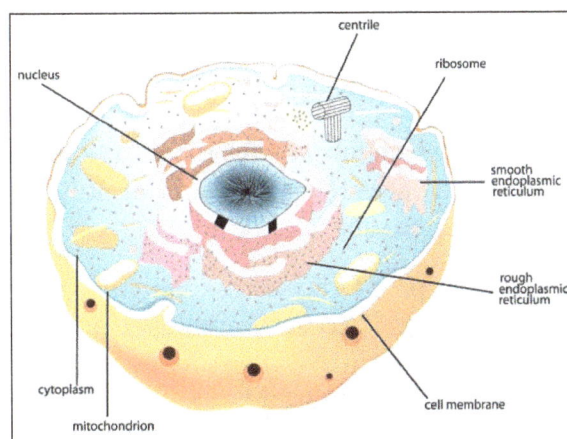

Cell and Nucleus

The nucleus is mechanically stable, possessing the ability to resist deformation. Additionally, the nucleus dynamically interacts with the surrounding cytoskeleton. The mechanical support and functional organization of the nucleus is contributed by several nuclear subcompartments, or nuclear bodies. These facilitate the various nuclear processes, particularly gene expression. These subcompartments together form the nucleoskeleton, and include the following.

Nuclear Membrane (Nuclear Envelope)

The Nuclear Lamina is a structure that is located near the inner nuclear membrane. It is highly proteinaceous and is fibrillar in appearance. The fibrillar nature of the lamina arises from the building blocks of the lamina; the laminin proteins. These first form dimers, and subsequently orientate in a head-to-tail fashion to produce filaments.

The nuclear skeletal network (karyoskeleton) that results serves as an anchor for chromatin. This maintains the shape of the nucleus spaces the nuclear pore complexes and organizes heterochromatin, the condensed region of DNA of nucleus. Additionally, it functions in nuclear processes such as DNA replication, particularly in the regulation of transcription factors.

The nuclear envelope itself has both an inner and outer nuclear lipid membranes. The outer nuclear membrane is continuous with the rough endoplasmic reticulum. Intermediate filaments form a network the nuclear lamina whilst other other inner nuclear membrane proteins mediate the interactions of the envelope with chromatin.

Nucleoskeleton

This is a network that provides support to the whole nucleus. It is made up of intermediate V type filaments composed of lamin proteins. It forms an elastic structure near the nuclear periphery and affords a viscoelastic property to the nucleus. At the nuclear envelope, the A and C type laminins provide a stiffness to the otherwise unstructured nucleus, whilst simultaneously supporting DNA processing events such as chromatin remodeling during replication.

B-type laminins provide support the processes of transcription and cellular signaling. There are also other structural proteins in the nucleoskeleton which impact the spatial organization of lamin proteins and other mechanical properties, such as resilience in response to stretch.

Nuclear Pore (Nuclear Pore Complex)

Nuclear pores are large protein complexes that form openings in the nuclear membrane. They sit at the point at which the inner nuclear membrane fuses with the outer nuclear membrane. They form aqueous channels that allow the selective movement of proteins, mRNA, tRNA, ribosome subunits and viruses in a bidirectional manner.

Molecules smaller than ~40 kDa can pass freely across the nuclear envelope and do not require passage through the NPC. Molecules >40 kDa each have a specific means of transportation across the nuclear envelope. Collectively, the system of transport is comprised of the protein cargo, nuclear transport receptors (NTRs) such as importins and exportins, and the small GTPase Ran.

The NTRs are regulated by Ran; it mediates the association of the NTRs with the cargo and provides the energy which drives the transport across the membrane. It is estimated that a typical mammalian cell has between 3000 and 4000 nuclear pores on its nuclear membrane and these pores have a diameter of approximately 600 A°.

Nucleolus

Nucleoli are small spherical bodies that are located in the nucleus; they are usually found in a centralized region but may be found close to the nuclear membrane. Nucleoli are distinct from the nuclear material, being built by the nucleolus organizing region (NOR) of a chromosome. The NORs store the genes required for full ribosomal synthesis - these genes specifically encode ribosomal RNA subunits.

There are approximately 10 NORs per nucleolus, and several chromosomes collaborate in nucleolus assembly, so that at most, cells contain an average of one or two. The presence of the nucleoli is determined by the cell identity; most animal and plant cells have a nucleolus. However, the presence of nucleoli is usually indicative of malfunction; for example, malignant, ageing or starving cells frequently display nucleoli of varying size, abundance and shape.

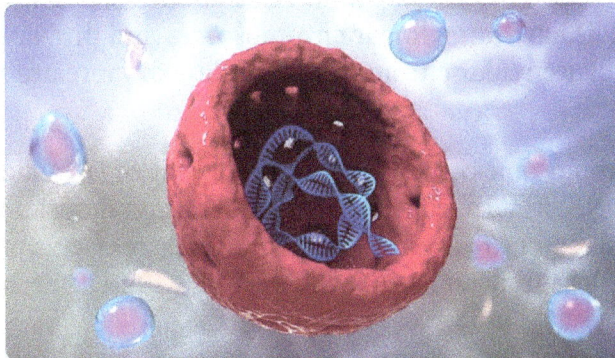

Nucleolus: Human body cell 3d illustration

In addition to DNA, the nucleolus is composed of two main regions – the pars fibrosa and the pars granulosa. The pars fibrosa contains proteins that participate in transcription and the pars granulosa houses the ribosomal precursors. Within the pars fibrosa, there is a fibrillar center which is distinct from a dense fibrillar component of the pars fibrosa.

The ribosomal RNA (rRNA) genes undergo transcription at the boundary of the two regions. Pre-rRNA processing follows this in the dense fibrillar component and continues in the granular component, where assembly of the rRNA with ribosomal proteins forms partially completed pre-ribosomal subunits ready for export to the cytoplasm.

A side from its function in ribosome biogenesis, the nucleolus is also the site of processing of several other noncoding RNAs. The nucleolus additionally plays a role in controlling the stability of the protein p53, a critical cell-cycle regulator.

The nucleolus disassembles during the prophase stage of mitosis; however, the dense fibrillar component remains associated with chromosomes and forms a secondary constriction point on the chromosome called the nucleolus-organizing region (NOR). This aids in genome organization during the subsequent interphase stage of the cell cycle.

Chromosome

A chromosome is a structure that occurs within cells and that contains the cell's genetic material. That genetic material, which determines how an organism develops, is a molecule of deoxyribonucleic acid (DNA). A molecule of DNA is a very long, coiled structure that contains many identifiable subunits known as genes. In prokaryotes, or cells without a nucleus, the chromosome is merely a circle of DNA. In eukaryotes, or cells with a distinct nucleus, chromosomes are much more complex in structure.

Structure of Chromosomes and Genes

A chromosome is an organized structure of DNA and protein that is found in cells. A chromosome is a single piece of coiled DNA containing many genes, regulatory elements and other nucleotide sequences. Chromosomes also contain DNA-bound proteins, which serve to package the DNA and control its functions. The word chromosome comes from the Greek chroma - color and soma - body due to their property of being very strongly stained by particular dyes. Chromosomes vary widely between different organisms. The DNA molecule may be circular or linear, and can be composed of 10,000 to 1,000,000,000 nucleotides in a long chain. Typically eukaryotic cells (cells with nuclei) have large linear chromosomes and prokaryotic cells (cells without defined nuclei) have smaller circular chromosomes, although there are many exceptions to this rule.

Today we know that a chromosome contains a single molecule of DNA along with several kinds of proteins. A molecule of DNA, in turn, consists of thousands and thousands of subunits, known as nucleotides, joined to each other in very long chains. A single molecule of DNA within a chromosome may be as long as 8.5 centimeters (3.3 inches). To fit within a chromosome, the DNA molecule has to be twisted and folded into a very complex shape.

Each chromosome has a constriction point called the centromere, which divides the chromosome into two sections, or "arms." The short arm of the chromosome is labeled the "p arm." The long arm of the chromosome is labeled the "q arm." The location of the centromere on each chromosome gives the chromosome its characteristic shape, and can be used to help describe the location of specific genes.

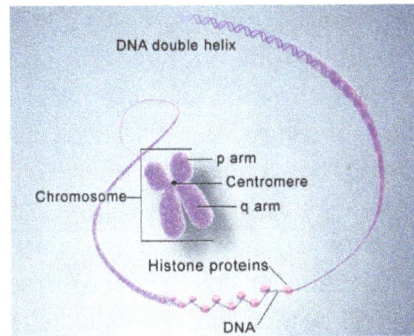

The arrangement of packets of genetic information in a chromosome is as follows.

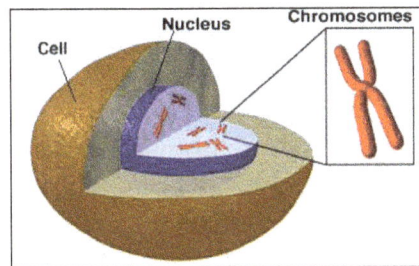

Furthermore, cells may contain more than one type of chromosome; for example, mitochondria in most eukaryotes and chloroplasts in plants have their own small chromosomes. The following are the different types of chromosomes.

Viral Chromosomes

The chromosomes of viruses are called viral chromosomes. They occur singly in a viral species and chemically may contain either DNA or RNA. The DNA containing viral chromosomes may be either of linear shape (e.g., T2, T3, T4, T5, bacteriophages) or circular shape (e.g., most animal viruses and certain bacteriophages). The RNA containing viral chromosomes are composed of a linear, single-stranded RNA molecule and occur in some animal viruses (e.g., poliomyelitis virus, influenza virus, etc.); most plant viruses, (e.g., tobacco mosaic virus, TMV) and some bacterio-phages. Both types of viral chromosomes are either tightly packed within the capsids of mature virus particles (virons) or occur freely inside the host cell.

Prokaryotic Chromosomes

The prokaryotes usually consist of a single giant and circular chromosome in each of their nucloids. Each prokaryotic chromosome consists of a single circular, double-stranded DNA molecule; but has no protein and RNA around the DNA molecule like eukaryotes. Different prokaryotic species have different sizes of chromosome.

Eukaryotic Chromosomes

The eukaryotic chromosomes differ from the prokaryotic chromosomes in morphology, chemical com-position and molecular structure. The eukaryotes (plants and animals) usually contain much more genetic informations than the viruses and prokaryotes, therefore, contain a great amount of genetic material, DNA molecule which here may not occur as a single unit, but, as many units called chromosomes. Different species of eukaryotes have different but always constant and characteristic number of chromosomes. In eukaryotes, nuclear chromosomes are packaged by proteins into a condensed structure called chromatin. This allows the very long DNA molecules to fit into the cell nucleus. The shape of the eukaryotic chromosomes is changeable from phase to phase in the continuous process of the cell growth and cell division. Chromosomes are the essential unit for cellular division and must be replicated, divided, and passed successfully to their daughter cells so as to ensure the genetic diversity and survival of their progeny. They are thin, coiled, elastic, contractile thread-like structures during the interphase (when no division of cell occurs) and are called chromatin threads which under low magnification look like a compact stainable mass, often called as chromatin substance or material. During metaphase stage of mitosis and prophase of meiosis, these chromatin threads become highly coiled and folded to form compact and individually distinct ribbon-shaped chromosomes. These chromosomes contain a clear zone called kinetochore or centromere along their length.

Eukaryotes (cells with nuclei such as plants, yeast, and animals) possess multiple large linear chromosomes contained in the cell's nucleus. Each chromosome has one centromere, with one or two arms projecting from the centromere, although, under most circumstances, these arms are not visible as such. In addition, most eukaryotes have a small circular mitochondrial genome, and some eukaryotes may have additional small circular or linear cytoplasmic chromosomes.

The number and position of centromeres is variable, but is definite in a specific chromosome of all the cells and in all the individuals of the same species. Thus, according to the number of the centromere the eukaryotic chromosomes may be acentric (without any centromere), mono centric (with one centromere), dicentric (with two centromeres) or polycentric (with more than two centromeres). The centromere has small granules or spherules and divides the chromosomes into two or more equal or unequal chromosomal arms.

According to the position of the centromere, the eukaryotic chromosomes may be rodshaped (telocentric and acrocentric), J-shaped (submetacentric) and V-shaped (metacentric) During the cell divisions the microtubules of the spindle are get attached with the chromosomal centromeres and move them towards the opposite poles of cell. Beside centromere, the chromosomes may bear terminal unipolar segments called telomeres. Certain chromosomes contain an additional specialized segment, the nucleolus organizer, which is associated with the nucleolus.

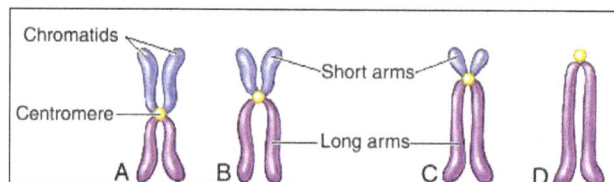

Position of the centromere in (A) metacentric; (B) submetacentric;
(C) acrocentric; and (D) telocentric chromosomes

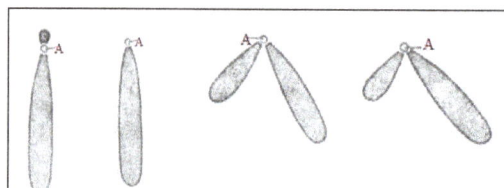

Centromere: Acrocentric, Telocentric, Metacentric, Sub metacentric

In the nuclear chromosomes of eukaryotes, the uncondensed DNA exists in a semi ordered structure, where it is wrapped around histones (structural proteins), forming a composite material called chromatin. Chromatin is the complex of DNA and protein found in the eukaryotic nucleus which packages chromosomes. The structure of chromatin varies significantly between different stages of the cell cycle, according to the requirements of the DNA.

Interphase Chromatin

During interphase (the period of the cell cycle where the cell is not dividing), two types of chromatin can be distinguished. The density of the chromatin that makes up each chromosome (that is, how tightly it is packed) varies along the length of the chromosome.

- Less dense regions are called euchromatin.

- Dense regions are called heterochromatin.

1. Euchromatin, which consists of DNA that is active, e.g., being expressed as protein.

2. Heterochromatin, which consists of mostly inactive DNA. It seems to serve structural purposes during the chromosomal stages. Heterochromatin can be further distinguished into two types:

- Constitutive heterochromatin, which is never expressed. It is located around the centromere and usually contains repetitive sequences.

- Facultative heterochromatin, which is sometimes expressed.

Individual chromosomes cannot be distinguished at this stage - they appear in the nucleus as a homogeneous tangled mix of DNA and protein.

Diploids and Haploids

In contrast to prokaryotes, most eukaryote are diploids, i.e., each somatic cell of them contains one set of chromosomes inherited from the maternal (female) parent and a comparable set of chromosomes (called homologous chromosomes) from the paternal (male) parent. The number of chromosomes in a dual set of a diploid somatic cell is called the diploid number (2n). The sex cells (sperms and ova) of a diploid eukaryote cell contain half the number of chromosomal sets found in the somatic cells and are known as haploid (n) cells. A haploid set of chromosome is also called genome. The fertilization process restores the diploid number of a diploid species.

Chemical Structure of Chromosomes

Chemically, the eukaryotic chromosomes are composed of deoxyribonucleic acid (DNA), ribonucleic acid (RNA), histone and non-histone proteins and certain metallic ions. The histone proteins have basic properties and have significant role in controlling or regulating the functions of chromosomal DNA. The non-histone proteins are mostly acidic and have been considered more important than histones as regulatory molecules. Some non-histone proteins also have enzymatic activities. The most important enzymatic proteins of chromosomes are phosphoproteins, DNA polymerase, RNA-polymerase, DPN-pyropbosphorylase, and nucleoside triphosphatase. The metal ions as Ca^+ and Mg^+ are supposed to maintain the oragnization of chromosomes intact.

Molecular Structure of Chromosomes

According to the recent and widely accepted theory of Dupraw and Hans Ris called unistranded theory, each eukaryotic chromosome is composed of a single, greatly elongated and highly folded nucleoprotein fibre of 100A o thick. This nucleo- protein fibre in its turn is composed of a single, linear, double stranded DNA molecule which remains wrapped in equal amounts of histone and non-histone proteins and variable amounts of different kinds of RNA. Dupraw produced a "folded-fibre Model" to show the ultrastructure of chromosome.

Fibre Folded Model

This model shows a highly folded nucleoprotein fibre in a chromosome and also suggests that how the nucleoprotein fibre of a chromosome replicates during cell division and how the nucleoprotein fibre of both chromatids remain held at the centromere by an unreplicated fibre segment to DNA until the anaphase.

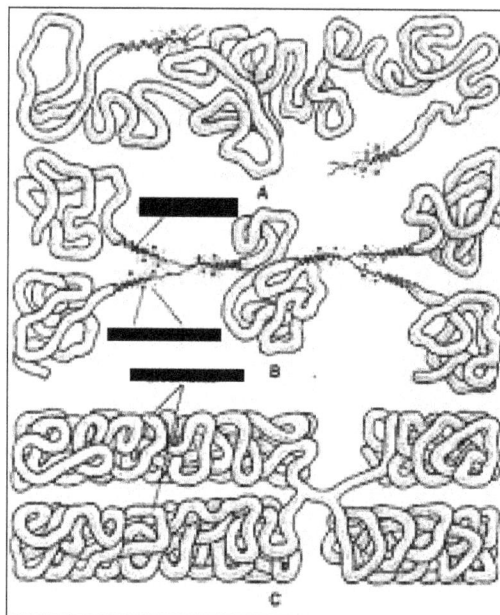

The folded fibre model of Dupraw for chromosomes in Interphase:
1. DNA molecule, 2.Protein molecules and Chromatids

Material of Chromosomes

The chromatin material of the eukaryotic chromosomes according to its percentage of DNA, RNA and proteins and consequently due to its, staining property has been classified into following by classical cytologists:

Euchromatin

The euchromatin is the extended form of chromatin and it forms the major, portion of chromosomes. The euchromatin has special affinity for basic stains and is genetically active because its component DNA molecule synthesizes RNA molecules only in the extended form of chromatin.

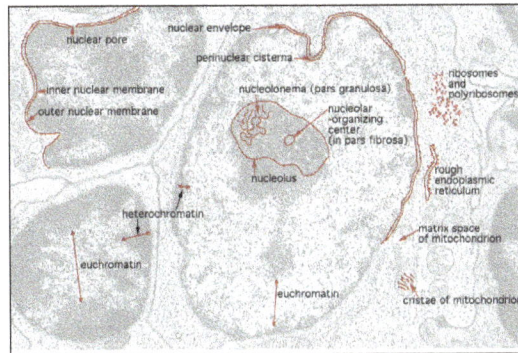

Heterochromatin

The heterochromatin is a condensed intercoiled state of chromatin, containing two to three times more DNA than euchromatin. However, it is genetically inert as it does not direct synthesize RNA (i.e., transcription) and protein and is often replicated at a different time from the rest of the DNA.

Recent molecular biological studies have identified three kinds of heterochromatins, namely constitutive, facultative and condensed heterochromatin. The constitutive heterochromatin is present at all times and in the nuclei of virtually all the cells of an organism. In a interphase nucleus, it tends to clump together to form chromocentre or false nucleoli. In Drosophila, for example, most pupal, larval and adult cells contain large blocks of constitutive heterochromatin that lie adjacent to centromeres. Constitutive heterochromatin contain highly repititive satellite DNA which is late replicating, it fails to replicate until late in the 5-phase and is then replicated during a brief period just before the G2. The facultative heterochromatin reflect the existence of a regulatory device designed to adjust the "dosages" of certain genes in the nucleus.

It is originated during the process called facultative heterochromatization: a process in which a chromosome or a set of chromosome becomes heterochromatic (turned off) in the cells of one sex, while remaining becomes heterochromatic (turned on) in the cells of opposite sex. In other words, it remains indirectly related to, sexual differentiation. The condensed heterochromatin is deeply staining tightly coiled chromatin which does not resemble with two other kinds of chromatin, has some specific role in gene regulation and is found in many interphase nuclei.

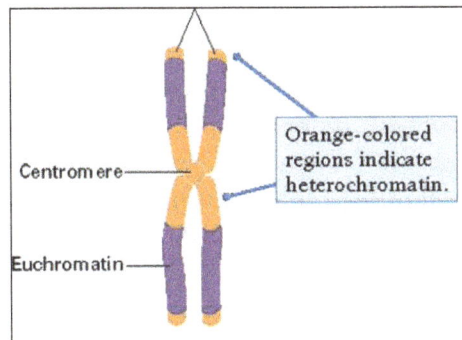

Kinds of Chromosomes

The eukaryotic chromosomes have been classified into autosomes and sex chromosomes. The autosomes have nothing to do with the determination of sex and exceed in number than sex chromosomes. The sex chromosomes determine the sex of their bearer. They are usually two in number and are usually of two kinds: X chromosomes and Y chromosomes.

Special types of Chromosomes

The eukaryotes besides possessing the usual type of chromosomes in their body cells contain some unusual and special types of chromosomes in some body cells or at some particular stage of their life cycle. The special types of eukaryotic chromosomes are following:

Polytene Chromosomes

The nuclei of the salivary gland cells of the larvae of dipterans like Drosophila have unusually long and wide chromosomes, 100 or 200 times in size of the chromosomes in meiosis and mitosis of the same species. This is particularly surprising, since the salivary gland cells do not divide after the glands are formed, yet their chromosomes replicate several times (a process called endomitosis) and become exceptionally giant-sized to be called polytene or multistranded chromosomes (discovered by Balbiani and named by Koller).The endomitosis process result in the production of 2X chromosomes, where X gives the number of multiplication cycle.

The polytene Chromosomes of the salivary gland cells of D. melanogaster contain 1000 to 2000 chromosomes, which are formed by nine or ten consecutive multiplication cycles and remain associated parallel to each other. Further, the polytene chromosomes have alternating dark and light bands along their length. The dark bands are comparable with the chromomeres of simple chromosomes and are disc-shaped structures occupying the whole diameter of chromosome. They contain euchromatin. The light bands or inter bands are fibrillar and composed of heterochromatin.

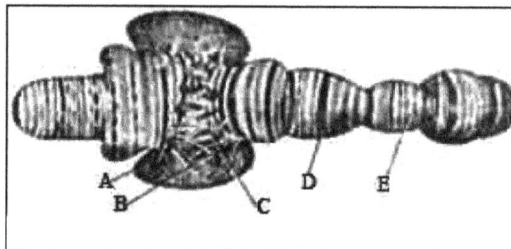

A. mRNA, B. Chromosome puff, C. Chromonemata, D. Dark band and E. Inter band

If the polytene chromosomes of dipteran larval salivary glands are examined at several stages of development; it is seen that specific areas (sets of bands) enlarge or "puff". Such puffs change location as development proceed, those at specific locations being correlated with particular developmental stages. This temporal puffing indicates changes in gene activity and involves several processes such as the accumulation of acidic proteins, despiralization of DNA, formation of chromonemal loops called Balbiani rings at the lateral sides of dark bands, synthesis of mRNA (messenger RNA) and storage (accumulation) of newly synthesized mRNA around the Balbiani rings.

Lampbrush Chromosomes

In diplotene stage of meiosis, the yolk rich oocytes of vertebrates contain the nuclei with many lamp brush shaped chromosomes of exceptionally large sizes. The lampbrush chromosomes are formed during the active synthesis of mRNA molecules for the future use by the egg during cleavage when no synthesis of mRNA molecules is possible due to active involvement of chromosomes in the mitotic cell division.

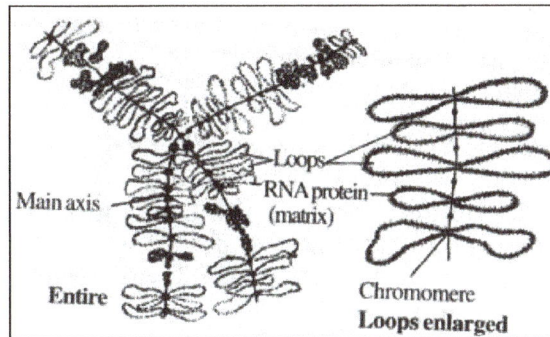

A lampbrush chromosome contains a main axis whose chromonemal fibres (DNA molecule) gives out lateral loops throughout its length. The loops produce the mRNA molecules of different kinds. In a mature egg, as the chromosome, contracts the lateral loops disappear.

B-chromosomes

Many plant (maize, etc.) and animal (such as insects and small mammals) species, besides having autosomes (A-chromosomes) and sex-chromosomes possess a special category of chromosomes called B-chromosomes without obvious genetic function. These Bchromosomes (also called supernumerary chromosomes, accessory chromosomes, accessory fragments, etc.) usually have a normal structure, are somewhat smaller than the autosomes and can be predominantly, heterochromatic (many insects, maize, etc.) or pro-dominantly euchromatic (rye).

In maize, their number per cell can vary from 0 to 30 and they adversely affect, development and fertility only when occur, in large amount. In animals, the B-chromosomes disappear from the non-reproductive (somatic) tissue and are maintained only in the cell-lines that lead to the reproductive organs. B-chromosomes have negative consequences for the organism, as they have deleterious effect because of abnormal crossing over during the meiosis of animals and abnormal nucleus divisions of the gametoophyte plants. In animals, Bchromosomes occur more frequently in females and the basis is non-disjunction. The non- disjunction of B-chromosomes of rye plant is found to be caused due to the presence of a heterochromatic knob at the end of long arm of B-chromosome.

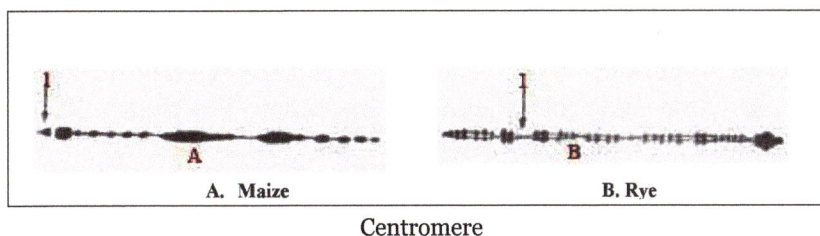

A. Maize   B. Rye

Centromere

The origin of the B-chromosomes is uncertain. In some animals they may be derivatives of sex chromosomes, but this is not the rule. They generally do not show any pairing affinity with the' A-chromosomes.

Holokinetic Chromosomes

The chromosomes of most plants and animals have centromeres that are situated at one specific position in each chromosome. In a number of animals, especially in insects of the order Hemiptera and a few, mostly monocotyledonous plants (Juncales, Cyperales), the kinetic activity is distributed over the entire chromosome and such chromosomes are called Holokinetic chromosomes. The term -diffuse centromere bas been used as an alternative but is not quite logical. In mitotic metaphase, the chromatids of a Holokinetic chromosome orient parallel in the equator: one chromatid towards one pole the other towards the other pole. This is also the way they separate in anaphase and they maintain this orientation until they arrive at the poles. Probably kinetic activity starts at one point and proceed from there on, orienting each unit to the preceding one.

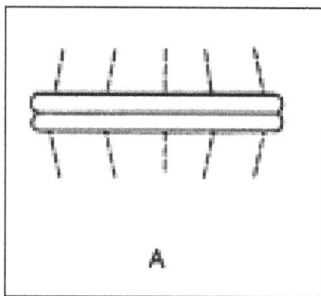

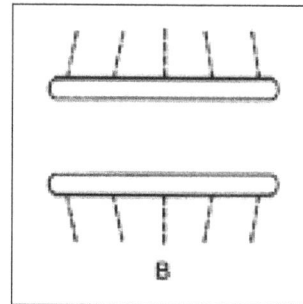

A. Holokinetic chromosome at mitotic metaphase   B. Holokinetic chromosome at mitotic anaphase

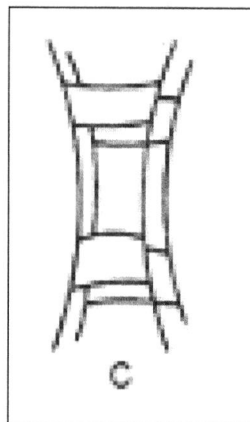

C. Meiotic metaphase I bivalent of a holokinetic chromosome pair

In a number of animals, especially in insects of the order Hemiptera and a few, mostly monocotyledonous plants (Juncales, Cyperales), the kinetic activity is distributed over the entire chromosome and such chromosomes ate caned Holokinetic chromosomes. The term -diffuse centromere bas been used as an alternative but is not quite logical. In 1966 Flach observed this type of centromere in some primitive DicotyledonS (Ranales: Myristica.Ascaris and pseudoscorpion Tityus also possess such polycentric chromosomes. In mitotic metaphase, the chromatids of a Holokinetic chromosome orient parallel in the equator: one chromatid towards one pole the other towards the other pole. This is also the way they separate in anaphase and they maintain this orientation until they arrive at the poles. Probably kinetic activity starts at one point and proceed from there on, orienting each unit to the preceding one.

Genetic Significance of Chromosomes

The chromosomes are considered as the organs of heredity because of following reasons:

1. They form the only link between two generations.

2. A diploid chromosome set consists of two morphologically similar (except the X and Y sex chromosomes) sets; one is derived from the mother and another from the father at fertilization.

3. The genetic material, DNA or RNA is localized in the chromosome and its contents are relatively constant from one generation to the next.

4. The chromosomes maintain and replicate the genetic informations contained in their DNA molecule and this information is transcribed at the right time in proper sequence into the specific types of RNA molecules which directs the synthesis of different types of proteins to form a body form like the parents.

Karyotype

A karyotype is the characteristic chromosome complement of a eukaryote species. The preparation and study of karyotypes is part of cytogenetics. The basic number of chromosomes in the somatic cells of an individual or a species is called the somatic number and is designated 2n. Thus, in humans 2n=46. In the germ-line (the sex cells) the chromosome number is n (humans: n=23). So, in normal diploid organisms, autosomal chromosomes are present in two copies. There may, or may not, be sex chromosomes. Polyploid cells have multiple copies of chromosomes and haploid cells have single copies. The study of whole sets of chromosomes is sometimes known as karyology. The chromosomes are depicted (by rearranging a microphotograph) in a standard format known as a karyogram or idiogram: in pairs, ordered by size and position of centromere for chromosomes of the same size. Karyotypes can be used for many purposes; such as, to study chromosomal aberrations, cellular function, taxonomic relationships, and to gather information about past evolutionary events.

Staining

The study of karyotypes is made possible by staining. Usually, a suitable dye is applied after cells have been arrested during cell division by a solution of colchicine. For humans, white blood cells are used most frequently because they are easily induced to divide and grow in tissue culture. Sometimes observations may be made on non-dividing (interphase) cells. The sex of an unborn fetus can be determined by observation of interphase cells.

Most (but not all) species have a standard karyotype. The normal human karyotypes contain 22 pairs of autosomal chromosomes and one pair of sex chromosomes. Normal karyotypes for females contain two X chromosomes and are denoted 46, XX; males have both an X and a Y chromosome denoted 46, XY. Any variation from the standard karyotype may lead to developmental abnormalities.

Six different characteristics of karyotypes are usually observed and compared:

1. Differences in absolute sizes of chromosomes. Chromosomes can vary in absolute size by as much as twenty-fold between genera of the same family: Lotus tenuis and Vicia faba

(legumes), both have six pairs of chromosomes (n=6) yet V. faba chromosomes are many times larger. This feature probably reflects different amounts of DNA duplication.

2. Differences in the position of centromeres. This is brought about by translocations.

3. Differences in relative size of chromosomes can only be caused by segmental interchange of unequal lengths.

4. Differences in basic number of chromosomes may occur due to successive unequal translocations which finally remove all the essential genetic material from a chromosome, permitting its loss without penalty to the organism (the dislocation hypothesis). Humans have one pair fewer chromosomes than the great apes, but the genes have been mostly translocated (added) to other chromosomes.

5. Differences in number and position of satellites, which (when they occur) are small bodies attached to a chromosome by a thin thread.

6. Differences in degree and distribution of heterochromatic regions. Heterochromatin stains darker than euchromatin, indicating tighter packing, and mainly consists of genetically inactive repetitive DNA sequences.

A full account of a karyotype may therefore include the number, type, shape and banding of the chromosomes, as well as other cytogenetic information. Variation is often found:

1. Between the sexes.

2. Between the germ-line and soma (between gametes and the rest of the body).

3. Between members of a population (chromosome polymorphism).

4. Geographical variation between races.

5. Mosaics or otherwise abnormal individuals.

Ideograms

Ideograms are a schematic representation of chromosomes. They show the relative size of the chromosomes and their banding patterns. A banding pattern appears when a tightly coiled chromosome is stained with specific chemical solutions and then viewed under a microscope. Some parts of the chromosome are stained (G-bands) while others refuse to adopt the dye (R-bands). The resulting alternating stained parts form a characteristic banding pattern which can be used to identify a chromosome. The bands can also be used to describe the location of genes or interspersed elements on a chromosome.

Below is an ideogram of each human chromosome. Next to the known schematic representation a chromosome was added that was rendered from DNA data. The G-bands, areas with proportional more A-T base pairs, are normally colored black in schematic representations. To compare the schematic ideograms the rendered chromosomes, were colored the A-T bases black and the G-C bases white. Blue areas in the rendered chromosomes identify bases not known yet.

The results are interesting. When comparing the schematic ideograms with the renderd chromosomes from our project, a significant conformancy can be seen. Most black areas on the left side show also black areas on the right side and white areas are also white on the "digital" chromosomes.

Ribosomes

A ribosome is a cell organelle. It functions as a micro-machine for making proteins. Ribosomes are organelles composed of ribosomal proteins (riboproteins) and ribonucleic acids (ribonucleoproteins). The word ribosome is made from taking 'ribo' from ribonucleic acid and adding it to 'soma', the latin word for body. Ribosomes are bound by a membrane but they are not membranous.

Ribosome: A Micro-machine for Manufacturing Proteins

A ribosome is basically a very complicated but elegant micro-'machine' for producing proteins. Each complete ribosome is constructed from two sub-units. A eukaryotic ribosome is composed of nucleic acids and about 80 proteins and has a molecular mass of about 4,200,000 Da. About two-thirds of this mass is composed of ribosomal RNA and one third of about 50+ different ribosomal proteins.

Ribosomes are found in prokaryotic and eukaryotic cells; in mitochondria, chloroplasts and bacteria. Those found in prokaryotes are generally smaller than those in eukaryotes. Ribosomes in mitochondria and chloroplasts are similar in size to those in bacteria. There are about 10 billion protein molecules in a mammalian cell and ribosomes produce most of them. A rapidly growing mammalian cell can contain about 10 million ribosomes. A single cell of *E. Coli contains* about 20,000 ribosomes and this accounts for about 25% of the total cell mass.

The proteins and nucleic acids that form the ribosome sub-units are made in the nucleolus and exported through nuclear pores into the cytoplasm. The two sub-units are unequal in size and exist in this state until required for use. The larger sub-unit is about twice as large as the smaller one.

The larger sub-unit has mainly a catalytic function; the smaller sub-unit mainly a decoding one. In the large sub-unit ribosomal RNA performs the function of an enzyme and is termed a ribozyme. The smaller unit links up with mRNA and then locks-on to a larger sub-unit. Once formed ribosomes are not static organelles. When production of a specific protein has finished the two sub-units separate and are then usually broken down. Ribosomes have only a temporary existence.

Sometimes ribosome sub-units admit mRNA as soon as the mRNA emerges from the nucleus. When many ribosomes do this the structure is called a polysome. Ribosomes can function in a 'free' state in the cytoplasm but they can also 'settle' on the endoplasmic reticulum to form 'rough endoplasmic reticulum'. Where there is rough endoplasmic reticulum the association between ribosome and endoplasmic reticulum (ER) facilitates the further processing and checking of newly made proteins by the ER.

Protein Factory: Site and Services

All factories need services such as gas, water, drainage and communications. For these to be provided there must a location or site.

Protein production also needs service requirements. A site requiring the provision of services is produced in a small ribosome sub-unit when a strand of mRNA enters through one selective cleft, and a strand of initiator tRNA through another. This action triggers the small sub-unit to lock-on to a ribosome large sub-unit to form a complete and active ribosome. The amazing process of protein production can now begin.

For translation and protein synthesis to take place many initiator and release chemicals are involved, and many reactions using enzymes take place. There are however general requirements and these have to be satisfied. The list below shows the main requirements and how they are provided:

- Requirement: A safe (contamination free) and suitable facility for the protein production process to take place.

- Provision: This facility is provided by the two ribosomal sub-units each of which is protected by a membrane covering. When the two sub-units lock together to form the complete ribosome, molecules entering and exiting can only do so through selective clefts or tunnels in the molecular structure.

- Requirement: A supply of information in a form that the ribosome can translate with a high degree of accuracy. The translation must be accurate in order that the correct proteins are produced.

- Provision: Information is supplied by the nucleus and delivered to the ribosome in the form of a strand of mRNA. When mRNA is formed in the nucleus introns (non-coding sections) are cut out, and exons (coding sections) are joined together by a process called splicing.

- Requirement: A supply of amino acids from which the ribosomal mechanism can obtain the specific amino acids needed.

- Provision: Amino acids, mainly supplied from food, are normally freely available in the cytoplasm.

- Requirement: A system that can select and lock-on to an amino acid in the cytoplasm and deliver it to the translation and synthesis site in the ribosome.

- Provision: Short strands of transfer ribonucleic acid (tRNA) made in the nucleus and available in the cytoplasm act as 'adaptor tools'. When a strand of tRNA has locked on to an amino acid the tRNA is said to be 'charged'. tRNA diffuses into the smaller ribosome sub-unit and each short tRNA strand will deliver ONE amino acid.

- Requirement: A means of releasing into the cytoplasm:

  a) A newly formed polypeptide,

  b) mRNA that has been used in the translating process, and

  c) tRNA that has delivered the amino acid it was carrying and is now 'uncharged'.

- Provision:

1. When a newly formed peptide chain is produced deep inside the ribosome large sub-unit, it is directed out to the cytoplasm along a tunnel or cleft.

2. 'Used' mRNA leaves the smaller ribosome sub-unit through a tunnel on the side opposite to its point of entry. Movement through the ribosome is brought about by a one-way only, intermittent movement of the ribosome along, and in the direction of, the incoming mRNA strand.

3. tRNA in the 'uncharged' state leaves via a tunnel in the molecular architecture of the ribosome large sub-unit.

In the ribosome there are Three Stages and Three operational Sites involved in the protein production line.

The three Stages are (1) Initiation, (2) Elongation and (3) Termination.

The three operational or binding SITES are A, P and E reading from the mRNA entry site (conventionally the right hand side).

Sites A and P span both the ribosome sub-units with a larger part residing in the ribosome large sub-unit, and a smaller part in the smaller sub-unit. Site E, the exit site, resides in the large ribosome sub-unit.

Table: Binding sites, positions and functions in a ribosome.

Binding Site	mRNA strand entry site	Biological term	Main processes
Site A	1st	Aminoacyl	Admission of codon of mRNA & 'charged' strand of tRNA. Checking and decoding and start of 'handing over' one amino acid molecule.
Site P	2nd	Peptidyl	Peptide synthesis, consolidation, elongation and transfer of peptide chain to site A.
Site E	3rd	Exit-to cytoplasm	Preparation of 'uncharged' tRNA for exit.

The Three stages:

1. Initiation: During this stage a small ribosome sub-unit links onto the 'start end' of an mRNA strand. 'Initiator tRNA' also enters the small sub-unit. This complex then joins onto a ribosome large sub-unit. At the beginning of the mRNA strand there is a 'start translating' message and a strand of tRNA 'charged' with one specific amino acid, enters site A of the ribosome. Production of a polypeptide has now been initiated.For the tRNA not to be rejected the three letter code group it carries (called an anti-codon) must match up with the three letter code group (called a codon) on the strand of mRNA already in the ribosome. This is a very important part of the translation process and it is surprising how few 'errors of translation' occur. [In general the particular amino acid it carries is determined by the three letter anticodon it bears, e.g. if the three letter code is CAG (Cytosine, Adenine, Guanine) then it will select and transport the amino acid Glutamine (Gln)].

2. Elongation: This term covers the period between initiation and termination and it is during this time that the main part of the designated protein is made. The process consists of a series of cycles, the total number of which is determined by the mRNA. One of the main events during elongation is translocation. This is when the ribosome moves along the mRNA by one codon notch and a new cycle starts.During the 'start-up' process the 'initiation tRNA' will have moved to site P and the ribosome will have admitted into site A, a new tRNA 'charged' with one amino acid.The 'charged' tRNA resides in site A until it has been checked and accepted (or rejected) and until the growing peptide chain attached to the tRNA in site P, has been transferred across by enzymes, to the 'charged' tRNA in site A. Here one new amino acid is donated by the tRNA and added to the peptide chain. By this process the peptide chain is increased in length by increments of one amino acid. [The peptide bond formation between the growing peptide chain and the newly admitted amino acid is assisted by peptidyl transferase and takes place in the large ribosome sub-unit. The reaction occurs between tRNA that carries the nascent peptide chain, peptidyl-tRNA and the tRNA that carries the incoming amino acid, the aminoacyl-tRNA]. When this has taken place the tRNA in site P, having transferred its peptide chain, and now without any attachments, is moved to site E the exit site.Next, the tRNA in site A, complete with a peptide chain increased in length by one amino acid, moves to site P. In site P riboproteins act to consolidate the bonding of the peptide chain to the newly added amino acid. If the peptide chain is long the oldest part will be moved out into the cytoplasm to be followed by the rest of the chain as it is produced. The next cycle with site A now empty translocation takes place. The ribosome moves on by a distance of one (three letter) codon notch along the mRNA to bring a new codon into the processing area. tRNA 'charged' with an attached amino acid now enters site A, and provided a satisfactory match of the mRNA codon and tRNA anti-codon is made, the cycle starts again. This process continues until a termination stage is reached.

3. Termination: When the ribosome reaches the end of the mRNA strand, a terminal or 'end of protein code' message is flagged up. This registers the end of production for the particular protein coded for by this strand of mRNA. 'Release factor' chemicals prevent any more amino acid additions, and the new protein (polypeptide) is completely moved out into the cytoplasm through a cleft in the large sub-unit. The two ribosome sub-units disengage, separate and are re-used or broken down.

Schematic of ribosome, for explanation see text.

Golgi Apparatus

Named after scientist Camillo Golgi, Golgi apparatus (Golgi complex) is membrane-bound sacs organelles that are involved in the modification (and synthesis) storage and transport of proteins and lipids.

Compared to other organelles, it has a unique appearance made up of pouches that are referred to as cisternae. Inside the cell, this organelle is located close to the endoplasmic reticulum (ER) near the nucleus of the cell.

Morphology

Golgi apparatus is made up of several flattened, stacked sacs referred to as cisternae. However, they can be highly pleomorphic, which means that they can change their shape for their function.

Depending on the type of cell, the number may vary from just a few to thousands. The cisternae are very small with a diameter ranging from 0.5 to 1.0 nm. Each of these is bound by a membrane and is held together by a matrix of proteins.

On the other hand, the entire Golgi is helped by cytoplasmic microtubules and have also been found to contain a number of important compartments, which include:

Cis and Trans Golgi Network

The Cis and trans are different faces of Golgi apparatus. The cis face, which is convex in appearance is closer to the endoplasmic reticulum and acts as the receiving compartment from the ER. On the opposite side is the trans face (also referred to as maturing face). Medial and trans cisternal compartments - they hold various molecules for a period of time.

Movement of Proteins through Golgi Apparatus

When proteins are produced in the ER, they have to pass through the Golgi apparatus for processing before being released into the cell to be used. Here, the protein released from the ER pass through the cis face to enter the Golgi apparatus. It is important for these molecules to pass through the cisternae stack so that they can be modified and packaged.

Different regions have different types of enzymes that act on these molecules to modify them. For instance, these enzymes may either add or remove sugar groups from the proteins thereby modifying them. Once they are appropriately modified, the protein molecules then move towards the trans face to be released into the cell cytoplasm.

Movement of Proteins through Golgi Cisternae

Several models have been proposed to help explain how proteins are transported through the cistern, these include:

1. Vesicular Transport Model: According to this model, cisternae are stable compartments through which the protein cargo move.

The protein cargo is transferred between the cisternae compartments with the help of vesicle carriers. The carriers transport the protein cargo through the compartments until they are released. Essentially, this process takes place from one compartment to another. That is, the carriers move the cargo proteins from one cisternae to the next.

At each compartment, these protein are processed by either adding or removing given groups (sugar, sulfate) and then moved to the next cisternae before ultimately arriving at the trans face to be released.

2. Cisternal Maturation Model: According to this model, the cisternae themselves move thereby transporting the protein cargo. Here, therefore, the protein cargo does not move (or moved by carriers). Rather, they remain intact in the compartment.

Enzymes then arrive to the compartment and convert the cis cisterna to medial (or medial to trans cisterna). This process sees the cargo being moved into the subsequent compartments as they have been converted until it reaches the last compartment and ultimately released in to the cell.

This process has been observed in yeast where live-cell video microscopy showed what appeared to be Golgi apparatus being converted to another thereby moving the cargo.

Protein Processing

Modification of protein starts in the ER where the following takes place:

- Addition of oligosaccharide (this is composed of 14 sugar residues).

- Removal of three glucose residues and a single mannose.

In the Golgi apparatus, the processing of proteins revolves around the modification of the carbohydrate group on the glycoproteins. The processing process involves changes to the N-linked oligosaccharides that were added while the protein was in the ER.

The process takes occurs as follows:

- Removal of additional mannose residues (3 mannose) on the glycoprotein.

- Sequentially adding N-acetylglucosamine.

- Removal of other mannose residues (2 or more).

- Addition of a fucose group and additional N-acetylglucosamines (2 or more).

- Addition of 3 galactose and 3 sialic acid molecules.

Processing of N-linked oligosaccharide of the lysosomal proteins is different from other proteins released in the plasma, this involves:

- Addition of N-acetylglucosamine phosphates to given mannose residues in the cis face.

- Removing the N-acetylglucosamine group and retention of mannose-6-phosphate. This process is aimed at phosphorylation of the mannose group, which makes it possible for the mannose-6-phosphate receptor to identify the molecule at the trans Golgi network.

- Modification of proteins also involves the addition to carbohydrates onto the side chains of acceptor serine and threonine residues of given amino acids in a process referred to as glycosylation.

This process also takes place in a sequential manner where single sugar residues are added. With some of these processes, the modification may also involve the addition to sulfate groups on the sugars. Glycosylation is of great importance in the modification process given that it can prevent the degradation of proteins.

It also serves to:

- Hold the protein in the endoplasmic reticulum until it has been properly folded.

- Direct the protein to the right organelle.

As mentioned, one of the main functions of the Golgi apparatus is to transport such molecules as proteins and lipids. However, this is only possible when these molecules are properly modified/processed.

Processing/modification of these molecules are of great significance given that it helps modify the molecule into the proper structure that will be identified at its destination. It is for this reason proteins are not released from the ER if they are not properly folded or from the Golgi apparatus if they are not properly modified.

In addition, additional groups are added onto these molecules to help in their transportation. For instance, a given functional group will be added onto a protein so that it can attach to a carrier and be transported to the right destination.

Chloroplasts

Chloroplasts are tiny plant powerhouses that capture light energy to produce the starches and sugars that fuel plant growth.

General Characteristics of Chloroplasts

The first photosynthetic eukaryotes originated more than 1000 million years ago through the primary acquisition of a cyanobacterial endosymbiont by a eukaryotic host, which subsequently gave rise to glaucophytes (whose photosynthetic organelles are called "cyanelles"), red algae (containing "rhodoplasts") and green algae and plants (with "chloroplasts"). Other major photosynthetic eukaryotic lineages arose when eukaryotic hosts engulfed a free-living photosynthetic eukaryote (e.g. red or green alga), initiating secondary and tertiary endosymbioses. Therefore, chloroplasts are organelles that are characteristic of plant and green algal cells, but still exhibit many prokaryotic features.

During evolution, the cyanobacterium-derived genome has undergone a dramatic reduction in size, mainly as a result of outright gene loss and the large-scale transfer of genes to the nuclear genome. Thus, the genomes of modern chloroplasts (plastomes) contain only 120-130 genes, most of which encode components of the organelle's gene expression machinery and its photosynthetic apparatus, and are organized in nucleoids that show both prokaryotic and eukaryotic features. However, chloroplasts contain many more protein species than their plastomes can code for. Hence, the majority of chloroplast proteins are now encoded by the nuclear genome and must be imported post-translationally into the organelle.

Apart from photosynthesis, chloroplasts are capable of performing many other specialized functions that are essential for plant growth and development — nitrate and sulphate assimilation, and the synthesis of amino acids and fatty acids, chlorophyll and carotenoids. To carry out these tasks, their membrane systems are equipped with specialized transport functions. The outer and inner envelope membranes mediate the import and sorting of proteins and the exchange of metabolites, while protein complexes in the thylakoid membranes implement the proton and electron transport processes that are an essential part of the photosynthetic light reactions. The thylakoids of land plants, where photosynthesis takes place, display an intricate architecture, with regions of stacked and appressed thylakoid membranes forming so-called grana. Moreover, plastids communicate with the nucleus by retrograde signaling to adjust the expression of nuclear genes according to the metabolic and developmental state of the organelle.

Chloroplast Structure

Nucleoids

A single mesophyll chloroplast can contain up to 300 chromosomes, which are organized into complex structures called "nucleoids", each consisting of 10-20 copies of the plastid genome, together with RNA and various proteins. Owing to their endosymbiotic origin and the fact that photosynthetic metabolism goes on all around them, nucleoids have a unique composition and organization, and display features typical of prokaryotic nucleoids, as well as attributes of eukaryotic chromatin. Nucleoids contain all the enzymes necessary for transcription, replication and segregation of the plastid genome. In addition, mRNA processing and editing, as well as ribosome assembly, take place in association with the nucleoid, suggesting that these processes occur co-transcriptionally. However, few nucleoid proteins have been characterized in detail.

Proteomic analysis of nucleoid preparations has identified new DNA-binding proteins, some of which were not inherited from the prokaryotic ancestors. One group of proteins in particular have been described, which contains a so-called SWIB domain that has previously been shown to be part of chromatin remodelling complexes in yeast. This domain is present in 20 proteins in Arabidopsis, and at least four of these are located in the chloroplast. The SWIB-domain proteins in chloroplasts are small proteins with a high isoelectric point and a high lysine content and might serve as functional replacements for the bacterial histone-like, DNA-binding HU proteins known from *Escherichia coli*. Thus one of them, SWIB-4, has a histone H1-like motif and binds to DNA, and recombinant SWIB-4 has been shown to induce compaction and condensation of nucleoids, and functionally complements an *E. coli* mutant that lacks the histone-like nucleoid structuring protein H-NS.

The two suppressor of variegation 4 (SVR4) proteins (SVR4 and SVR4-like), originally identified in Arabidopsis, both have orthologues in all dicot and monocot plants sequenced so far, whereas the moss Physcomitrella patens and spikemoss Selaginella moellendorffii contain only one gene copy, indicating that a gene duplication took place in the progenitor of vascular plants. Inactivation of either SVR4 or SVR4-like in Arabidopsis results in seedling lethality. Both proteins are localized in the chloroplast, expressed during early stages of chloroplast development, and contain 20% negatively charged amino acid residues Given the inherent risk of random aggregation of the negatively charged nucleic acids and basic proteins, such as histones and ribosomal subunits, SVR4 and SVR4-like could function as negatively charged molecular chaperones that mimic nucleic acids or serve as decoys to allow for the establishment of productive DNA/RNA-protein interaction in developing chloroplasts, where dramatic rearrangements in nucleoid organization take place.

Thylakoid Architecture

A structural hallmark of thylakoid membranes in plants are the so-called "grana". Grana cylinders are made up of stacks of flat grana membrane discs with a diameter of about 300-600 nm, which are enwrapped in (and interconnected by) the unstacked stroma lamellae. Tightly curved margins form the periphery of each discoid sac. For a typical granum from *Arabidopsis thaliana* the membrane bilayers are on average 4.0 nm thick, lumen thickness is 4.7 nm and discs are separated by a 3.6 nm gap.

The exact three-dimensional architecture of grana is still under debate, and two quite different types of models have been proposed: the "helical" model and several "fork/bifurcation" models. In the helical model, thylakoids comprise a fretwork of stroma lamellae, which wind around grana stacks as a right-handed helix, connecting individual grana disks *via* narrow membrane protrusions The grana are connected to each other solely by the stroma lamella helices, which are tilted at an angle ranging from 10 to 25°, with respect to the grana stacks and make multiple contacts with successive layers in the grana through slits located in the rims of the stacked discs. The fork/bifurcation models, on the other hand, postulate that the grana themselves are formed by bifurcations of stroma lamellae. Thus, Arvidsson and Sundby suggested that a granum contains piles of repeat units, each containing three grana discs, which are formed by symmetrical invaginations of a thylakoid pair caused by the bifurcation of the thylakoid membrane. Shimoni *et al* presented another model in which grana discs are paired units formed by simple bifurcation of stroma thylakoids. Here, the granum-stroma assembly is formed by bifurcations of the stroma lamellar membranes into multiple parallel discs. The stromal membranes form wide lamellar sheets that intersect the granum body roughly perpendicular to the long axis of the granum cylinder. In this model, adjacent granum layers are joined not only through the stroma lamellae, but *via* the bifurcations and through direct membrane bridges. This model can also be used to explain the rearrangements seen in thylakoids during state transitions. The mutual incompatibility of the helical and bifurcation models has led to a great deal of debate, but recent tomographic data clearly support the helical model.

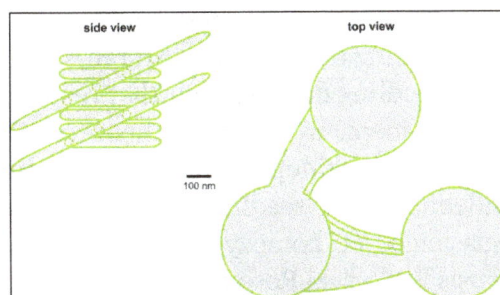

The helical model of thylakoid architecture

In figure, a fretwork of stroma lamellae, which wind around the ascending grana stacks as a right-handed helix connects to individual grana discs *via* narrow membrane protrusions (indicated by dotted circles in the side view).

Lateral Heterogeneity of Thylakoids

The term "lateral heterogeneity" refers to the observation that grana and stroma lamellae differ in their protein composition. Photosystem II and light-harvesting complex (LHC) II are concentrated in the grana, while photosystem I with its LHCI and the chloroplast ATP synthase are localized in the unstacked thylakoid regions, that is the stroma lamellae and grana end membranes. The cytochrome $b_6 f$ complex (Cyt $b_6 f$) can be found in both appressed and non-appressed regions of thylakoids. The NDH complex and the PGRL1-PGR5 heterodimer – the two thylakoid complexes specifically involved in cyclic electron flow – are less abundant than the aforementioned four major thylakoid complexes and are located in the stroma lamellae, where they can functionally interact with photosystem I as an electron donor. While the bulkiness of the NDH complex precludes its location in grana, PGRL1 homodimers have been detected in grana.

Detection of several of the major thylakoid multiprotein complexes in margin-enriched fractions of thylakoids by biochemical methods has been reported in some experiments. However, the marked curvature of thylakoid membranes at the grana margins is essentially incompatible with the presence of the larger multiprotein complexes at these sites. Therefore, grana margins have been thought to be essentially protein-free. However, following the recent demonstration, by immunogold labelling, that CURT1 proteins — small polypeptides with two transmembrane regions and a putative N-terminal amphipathic helix — are localized to grana margins this view must be revised. Interestingly, the CURT1 proteins appear to control the level of grana stacking, which points to an unsuspected role of grana margins in regulating the fraction of thylakoid membranes incorporated into the appressed regions that make up grana.

Chloroplast Functions

Chloroplast Proteins: Mutants and Proteomes

Estimates for the size of the chloroplast proteome in Arabidopsis range from 2000 to 4400 different proteins. In the course of the Chloroplast 2010 Project, homozygous mutants for several thousand nuclear genes with chloroplast functions were identified and phenotypically characterized. Despite extensive screening, for several hundred genes no homozygous mutant alleles were discovered, suggesting that these might represent genes with essential functions. More recently, lines that had failed to yield any homozygotes when grown in soil were tested for homozygous lethality owing to defects either in seed or seedling development. Mutants arrested at various stages of seed development (and with defects in seedling development that responded to supplementation with sucrose, amino acids or to CO_2 enrichment) were indeed uncovered. This resulted in an annotation of more than 200 publically available Arabidopsis mutants, including 36 and 33 genes with one and two, respectively, independent seed- or seedling-development-defective mutant alleles. The study also resulted in the submission of 521 homozygous mutants and 128 seed stocks segregating for lethal alleles to the Arabidopsis Biological Resource Center.

Proteomics usefully complement the reverse genetics approach to chloroplast function outlined above. Driven by recent advances in bioanalytical and computational technologies, the strategy allows for identification, and reasonably accurate quantification, of thousands of proteins in complex mixtures, as well as the ability to characterize post-translational modifications, such as acetylation, glycosylation and phosphorylation. An attempt to obtain a high-quality inventory of the plastid proteome has led to the identification of 1564 and 1559 proteins for maize and Arabidopsis, respectively. These estimates were based on both manual curation of published experimental information, including more than 150 proteomics studies devoted to different subcellular fractions, and new quantitative proteomics experiments on plastid subfractions. These figures correspond to an estimated 40% and 50% of all plastid proteins in maize and Arabidopsis, respectively — the most comprehensive inventory assembled so far.

Chlorophyll Biosynthesis

Biosynthesis of chlorophyll takes place in the plastid, and the initial steps in the pathway leading to protoporphyrin IX are common to the biosynthesis of other tetrapyrroles, such as heme. Important discoveries in chlorophyll biosynthesis include the demonstration that plastid glutamyl-transfer RNA is involved in the formation of glutamate-1-semialdehyde, which is subsequently converted into 5-aminolevulinic acid — the universal precursor of tetrapyrrole biosynthesis in all organisms, and the finding that the enzyme Mg-chelatase (which catalyses the insertion of the Mg^{2+} ion into protoporphyrin IX) contains three different protein subunits: ChlH, ChlI, and ChlD. Later, the GENOME UNCOUPLER4 (GUN4) was found to bind both the substrate and the product of the Mg-chelatase, thereby dramatically enhancing the activity of the enzyme. GUN4 also reduces the threshold Mg^{2+} concentration required for activity at low porphyrin concentrations, and it was proposed to have a protective function in tetrapyrrole trafficking and to control Mg-chelatase activity at physiologically significant Mg^{2+} concentrations. More recently, it was shown that GUN4 interacts with the ChlH subunit of the enzyme.

One of the least understood steps in chlorophyll biosynthesis is the formation of the isocyclic "fifth" ring (ring-E), which is catalysed in plants by the aerobic cyclase system (ACS). The overall cyclase reaction is a six-electron oxidation proposed to occur in three sequential steps:

(a) Hydroxylation of the methyl-esterified ring-C propionate by incorporation of atmospheric oxygen;

(b) Oxidation of the resulting alcohol to the corresponding ketone;

(c) Reaction of the activated methylene group with the γ -mesocarbon of the porphyrin nucleus in an oxidative reaction involving the removal of two protons to yield the "fifth" ring.

At the biochemical level, the ACS requires both soluble and membrane-bound chloroplast fractions and, in barley, at least two mutants exist (xantha-l and viridis-k) which are defective in the membrane components. Thus, the ACS may be composed of three gene products: A soluble protein and two membrane-bound components — one encoded in barley by Xantha-l and the other by Viridis-k. So far, only AcsF (which corresponds to Xantha-l in barley, CRD1 in Chlamydomonas or CHL27 in Arabidopsis) has been unambiguously identified. As diiron enzymes are known to perform hydroxylation and cyclization of keto intermediates, AcsF could be involved in one or more of the proposed cyclase steps. Recent progress has come from pull-down experiments using

FLAG-tagged versions of the two AcsF-like gene products in Synechocystis in combination with protein mass spectrometry, which have identified the soluble YCF54 protein as a new putative subunit of ACS. Inactivation of the Synechocystis ycf54 gene resulted in significantly reduced chlorophyll levels, marked accumulation of the substrate of the cyclase, Mg-protoporphyrin IX methyl ester, and only traces of its product, protochlorophyllide, indicating that YCF54 is essential for the activity and stability of the oxidative cyclase. Future experiments must clarify whether YCF54 is the long-sought soluble component of the cyclase system, or whether it functions in AcsF synthesis/maturation or in cyclase assembly. Low chlorophyll accumulation A (LCAA), the tobacco homologue of YCF54, might have an additional role in the feedback-control of 5-aminolevulinic acid biosynthesis. Because the structure of YCF54 is similar to that of the photosystem II assembly factor Psb28, YCF54 might also be involved in coordinating chlorophyll biosynthesis and photosystem biogenesis.

Table: Accessory factors involved in the assembly of thylakoid multiprotein complexes in plants and cyanobacteria.

PSII	PSI	Cyt b_6f	cpATPase	NDH
HCF136[A]/YCF48[S]	YCF3[C,T]	CCS1[C]	ALB4[A]	AtCYP20-2[A]
ALB3[A,C]/Slr1471[S]	YCF4[C,T]	CCB1[C,A]	AtCGL160[A]	CRR1[A]
YCF39[S]	YCF37[S]/Pyg7	CCB2[C,A]		CRR6[A]/Slr1097[S]
LPA1[A]/REP27[C]	PPD1[T]	CCB4[C,A]		CRR7[A]
LPA2[A]	Y3IP1[T,A]	DAC[A]		CRR41[A]
LPA3[A]	PBF1[T]			CRR42[A]
Slr2013[S]	HCF101[A]			NDF5[A]
Psb27[S]/ LPA19[A]	RubA[S]			PAM68L[A]
Psb28[S]				
Psb29/THF1[S,A]				
PratA[S]				
Pitt[S]				
AtCYP38[A]				
PAM68[A,S]				

Where, Organism in which the assembly factor was functionally characterized: A, Arabidopsis, C, Chlamydomonas, S, Synechocystis or Synechococcus, T, tobacco. Abbreviations: cpATPase, chloroplast ATP synthase; Cyt, cytochrome; PSI, photosystem I; PSII, photosystem II.

Photosynthesis: New Proteins and New Functions

It comes as a surprise to learn that some proteins that are directly involved in the light reactions of photosynthesis have remained unidentified until very recently. Thus, although antimycin A-sensitive cyclic electron flow (AA-sensitive CEF), which serves to recycle electrons from ferredoxin to plastoquinone, was discovered by Arnon and co-workers more than 50 years ago, it is only a few years since the proteins responsible were identified. A role in AA-sensitive CEF has been attributed to the two thylakoid proteins PGR5 and PGRL1 ever since their identification, but this assignment has remained controversial. Indeed, current technical limitations still preclude unequivocal clarification of their precise function in CEF in vivo, but recent biochemical experiments have shown that PGRL1/PGR5 complexes possess ferredoxin-plastoquinone reductase (FQR) activity in vitro. Consequently, PGRL1-PGR5 complexes in flowering plants appear to shuttle between photosystem I and the cytochrome (Cyt) b_6f complex, whereas in the green alga Chlamydomonas PGRL1 (but not PGR5) has been detected in a photosystem I cytochrome b_6f supercomplex that has intrinsic CEF activity.

The second pathway mediating CEF involves the so-called "NAD(P)H dehydrogenase complex" or "NDH complex". Although the plant NDH complex is related to the NADH dehydrogenase complexes of bacteria and mitochondria, its function and composition are enigmatic. Recently, the Shikanai group has identified three novel subunits of plant NDH (CRR-31, -J and -L) and their functional characterization clearly indicated that CRR-31 supplies a docking site for ferredoxin . Therefore it can be concluded that the plant NDH complex accepts electrons from ferredoxin rather than NAD(P)H. Consequently, the authors of the first study proposed that the term "NDH" be retained, but used to mean "NADH dehydrogenase-like complex" rather than "NAD(P)H dehydrogenase complex". In a strict sense, the NDH complex is also an FQR like the PGRL1/PGR5 complex. With respect to its physical interaction with other thylakoid complexes, the NDH complex has been shown, on the basis of genetic and biochemical experiments , as well as by electron microscopy analyses, to form super-complexes with photosystem I, such that two photosystem I complexes bind to one NDH complex, with Lhca5 and 6 acting as linkers . Interestingly, the association of photosystem I with light-harvesting 1 proteins and the NDH complex had evolved before the emergence of vascular plants, as evidenced by analyses of photosystem I in the moss *P. patens*.

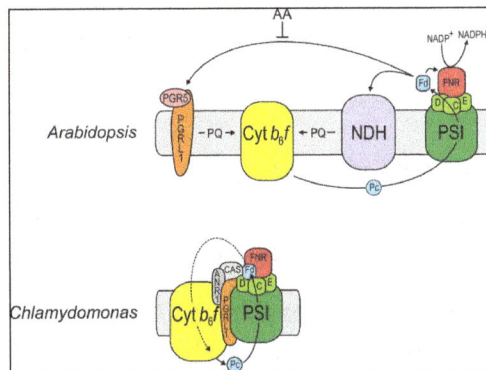

Model for the different roles of PGRL1 in cyclic electron flow in vascular plants and green algae

In figure, vascular plant Arabidopsis, cyclic electron flow (CEF) around photosystem I (PSI) operates via two partially redundant pathways, an NDH-dependent and the PGRL1/PGR5-dependent pathway. Only the latter is inhibited by AA. Note that both the PGRL1/PGR5 and the NDH complex (via Lhca5 and 6) can physically interact with PSI in Arabidopsis and accept electrons only from ferredoxin (Fd). In the green alga Chlamydomonas, CEF can be mediated by a PSI-Cyt b_6f-PGRL1-ANR1-CAS supercomplex. FNR, ferredoxin NADP+ oxidoreductase; Pc, plastocyanin; PQ, plastoquinone; C, D and E, stromal subunits of PSI.

The major thylakoid protein complexes require not only structural proteins that serve as subunits but also accessory proteins that mediate the correct assembly of the complexes. Here, a plethora of factors have been identified during the last few years and a picture emerged in which such accessory factors constitute an integrative network mediating the stepwise assembly of multiprotein complex components. Thus, it appears that besides the set of actual subunits of the photosystems, even more proteins are required for the expression of the chloroplast-encoded subunits and their assembly. In Table, we provide a list of the current inventory of assembly factors identified in Arabidopsis, Chlamydomonas, tobacco or cyanobacteria (Synechocystis or Synechococcus). Interestingly, two photosystem I assembly factors are encoded by chloroplast genes: ycf3 and ycf4. An important function in the chloroplast protein import machinery has been recently assigned to another chloroplast open reading frame, ycf1. However, such a tentative assignment of an important function of the encoded Tic214 protein is somehow at variance with the observation that chloroplast genomes of Poaceae species lack the ycf1 gene.

Novel Retrograde Signals

The term "retrograde signalling" refers to the idea that signals emanating from chloroplasts or mitochondria can modulate nuclear gene expression. Proposed almost 30 years ago, the initial notion that a single plastid signal might regulate the expression of nuclear genes involved in plastid biogenesis has since expanded to accommodate the insight that multiple signals are produced by plastids. While the ultimate effects of retrograde signalling on nuclear gene expression have now been clearly defined, many aspects of the initiation and transmission of the signals, and their mode of action, remain unresolved, speculative or controversial. Relevant signals are thought to be derived from various sources, including (a) the pool of reactive oxygen species (ROS), (b) the reduction/oxidation (redox) state of the organelle, (c) organellar gene expression, and (d) the tetrapyrrole pathway. More recently, "brand-new" retrograde signaling pathways have been described that involve (e) metabolites — particularly 3'-phosphoadenosine 5'-phosphate (PAP) and methylerythritol cyclodiphosphate (MEcPP) - and (f) a carotenoid derivative (β-cyclocitral [β-CC]).

Synthetic Biology

Synthetic biology can broadly be defined as "the deliberate (re)design and construction of novel biological and biologically based systems to perform new functions for useful purposes, that draws on principles elucidated from biology and engineering."

The plastome, at least in some species, such as tobacco, tomato and Chlamydomonas, can be manipulated by genetic transformation with large constructs made up of foreign or synthetic DNA segments. In fact, due to its prokaryotic origin, the chloroplast genome offers many advantages for genetic engineering because its genes are organized in operons and many are co-expressed from a single promoter as a polycistronic transcript that may subsequently be processed further into monocistronic mRNAs. Moreover, no position effects or epigenetic gene-silencing mechanisms, like those observed with nuclear transgenes, have been reported in chloroplasts. These features make the chloroplast compartment especially amenable to the application of synthetic biology to goals such as the sustainable synthesis of chemicals and high-value products. Lu *et al.* successfully demonstrated this by expressing the tocochromanol pathway (which produces tocopherols and tocotrienols, collectively called "vitamin E") in the chloroplasts of tobacco and tomato and achieving up to a tenfold increase in total tocochromanol accumulation. This represents a prime example of how overexpression of enzymes in the chloroplast can redirect photosynthetically generated carbon skeletons from the endogenous isoprenoid biosynthetic pathway into the production of higher levels of tocopherols and tocotrienols.

It is highly desirable that novel pathways introduced into the chloroplast should be able to tap directly the chemical energy derived from sunlight in the form of ATP, NADPH or even photo-re-duced ferredoxin. One group of enzymes which could potentially be used for this purpose are the cytochrome P450 mono-oxygenases (P450s), which are represented in all biological kingdoms and constitute one of the largest superfamilies of enzymes known. Most P450s are located in the endoplasmatic reticulum, where they act as key enzymes in the biosynthesis of a large number of high-value bioactive natural compounds. Many of these compounds are normally made in very small quantities and are difficult to produce by chemical synthesis due to their often complex structures. P450s generally obtain the electrons needed for their catalytic reactions from NADPH or NADH, but bacterial and mitochondrial P450s are also known to accept electrons from ferre-doxin. Therefore, a direct link between photoreduced ferredoxin and P450s is possible if the evo-lutionary compartmentalization of the photosystems in the chloroplasts and of the majority of the P450 pathways in the endoplasmatic reticulum can be broken down.

The potential value of combining P450-mediated mono-oxygenation reactions with photosynthe-sis was first demonstrated *in vitro* when spinach chloroplasts were brought together with micro-somes from yeast expressing a fusion between a P450 from rat (CYP1A1) and a reductase. This mixture supported the light-driven conversion of the P450 substrate 7-ethoxycoumarin into 7-hy-droxycoumarin. More recently, it was shown *in vitro* that electrons supplied by photosystem I purified from barley could be transferred with high efficiency to a P450 (CYP79A1) from *Sorghum bicolor via* ferredoxin, thus eliminating the need for an NADPH recycling system and a reductase. Subsequently, it was shown that the P450-catalysed pathway for the biosynthesis of dhurrin (a cyanogenic glycoside) can be transferred from the cytosolic endoplasmatic reticulum of *S. bicolor* into the tobacco chloroplast. To this end, fusion proteins between a chloroplast transit peptide and the coding regions of two P450 enzymes and a uridine 5'-diphosphate (UDP) glucosyltransferase, which together constitute the route to dhurrin biosynthesis, were successfully expressed in the chloroplasts of transiently transformed tobacco leaves. Interestingly, the chloroplast was able to provide the heme cofactor for the proper assembly of the P450s, the tyrosine and UDP-glucose substrates. The electron-demanding P450-catalysed synthesis of dhurrin was driven by directly tapping into light-driven reduction of ferredoxin by photosystem I. Thus, this example demon-strates that P450s that normally reside in the endoplasmic reticulum membranes can be targeted to the chloroplast and inserted into the thylakoids and can act as receptors for electrons from the light reactions of photosynthesis for use in the biosynthesis of dhurrin.

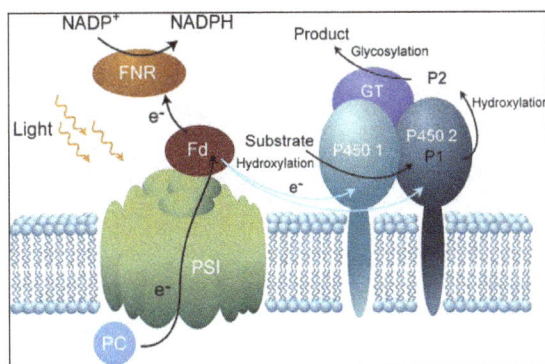

Schematic representation of a light-driven metabolon introduced into the thylakoids

In figure, Photosystem I (PSI) receives electrons from photosystem II *via* plastocyanin (PC) and directs them to ferredoxin (Fd), which give them either to the ferredoxin NADP+ oxidoreductase

(FNR) for NADPH production or directly to the two P450 enzymes (P450s). The two membrane-bound P450s hydroxylate the substrate in two consecutive steps, and this is followed by glycosylation by a soluble glucosyltransferase (GT) to form the final stable product. The novel aspect of this approach is that photosynthetic reducing power, in the form of reduced ferredoxin, is used directly by a novel biosynthetic pathway to produce the product without the need for numerous energy consuming metabolic conversions.

Endoplasmic Reticulum

Endoplasmic reticulum (ER), in biology, is a continuous membrane system that forms a series of flattened sacs within the cytoplasm of eukaryotic cells and serves multiple functions, being important particularly in the synthesis, folding, modification, and transport of proteins. All eukaryotic cells contain an endoplasmic reticulum (ER). In animal cells, the ER usually constitutes more than half of the membranous content of the cell. Differences in certain physical and functional characteristics distinguish the two types of ER, known as rough ER and smooth ER.

Rough ER is named for its rough appearance, which is due to the ribosomes attached to its outer (cytoplasmic) surface. Rough ER lies immediately adjacent to the cell nucleus, and its membrane is continuous with the outer membrane of the nuclear envelope. The ribosomes on rough ER specialize in the synthesis of proteins that possess a signal sequence that directs them specifically to the ER for processing. (A number of other proteins in a cell, including those destined for the nucleus and mitochondria, are targeted for synthesis on free ribosomes, or those not attached to the ER membrane.) Proteins synthesized by the rough ER have specific final destinations. Some proteins, for example, remain within the ER, whereas others are sent to the Golgi apparatus, which lies next to the ER. Proteins secreted from the Golgi apparatus are directed to lysosomes or to the cell membrane; still others are destined for secretion to the cell exterior. Proteins targeted for transport to the Golgi apparatus are transferred from ribosomes on rough ER into the rough ER lumen, which serves as the site of protein folding, modification, and assembly.

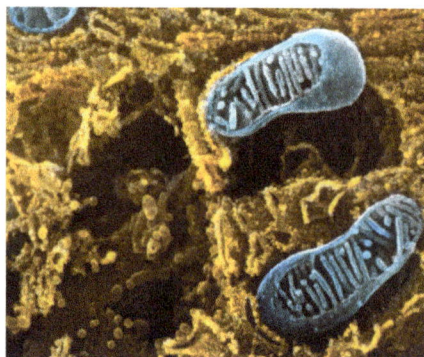

Endoplasmic Reticulum; Organelle: A scanning electron micrograph of a pancreatic acinar cell, showing mitochondria (blue), rough endoplasmic reticulum (yellow; ribosomes appear as small dots), and Golgi apparatus (gray, at centre and lower left).

The proximity of the rough ER to the cell nucleus gives the ER unique control over protein processing. The rough ER is able to rapidly send signals to the nucleus when problems in protein synthesis and folding occur and thereby influences the overall rate of protein translation. When misfolded or unfolded proteins accumulate in the ER lumen, a signaling mechanism known as the unfolded protein response (UPR) is activated. The response is adaptive, such that UPR activation triggers reductions in protein synthesis and enhancements in ER protein-folding capacity and ER-associated protein degradation. If the adaptive response fails, cells are directed to undergo apoptosis (programmed cell death).

Smooth ER, by contrast, is not associated with ribosomes, and its functions differ. The smooth ER is involved in the synthesis of lipids, including cholesterol and phospholipids, which are used in the production of new cellular membrane. In certain cell types, smooth ER plays an important role in the synthesis of steroid hormones from cholesterol. In cells of the liver, it contributes to the detoxification of drugs and harmful chemicals. The sarcoplasmic reticulum is a specialized type of smooth ER that regulates the calcium ion concentration in the cytoplasm of striated muscle cells.

The highly convoluted and labyrinthine structure of the ER led to its description in 1945 as a "lace-like reticulum" by cell biologists Keith Porter, Albert Claude, and Ernest Fullman, who produced the first electron micrograph of a cell. In the late 1940s and early 1950s, Porter and colleagues Helen P. Thompson and Frances Kallman introduced the term endoplasmic reticulum to describe the organelle. Porter later worked with Romanian-born American cell biologist George E. Palade to elucidate key characteristics of the ER.

Mitochondria

Mitochondria (singular: mitochondrion) are organelles within eukaryotic cells that produce adenosine triphosphate (ATP), the main energy molecule used by the cell. For this reason, the mitochondrion is sometimes referred to as "the powerhouse of the cell". Mitochondria are found in all eukaryotes, which are all living things that are not bacteria or archaea. It is thought that mitochondria arose from once free-living bacteria that were incorporated into cells.

Function of Mitochondria

Mitochondria produce ATP through process of cellular respiration-specifically, aerobic respiration, which requires oxygen. The citric acid cycle, or Krebs cycle, takes place in the mitochondria. This cycle involves the oxidation of pyruvate, which comes from glucose, to form the molecule acetyl-CoA. Acetyl-CoA is in turn oxidized and ATP is produced.

The citric acid cycle reduces nicotinamide adenine dinucleotide (NAD^+) to NADH. NADH is then used in the process of oxidative phosphorylation, which also takes place in the mitochondria. Electrons from NADH travel through protein complexes that are embedded in the inner membrane of the mitochondria. This set of proteins is called an electron transport chain. Energy from the electron transport chain is then used to transport proteins back across the membrane, which power ATP synthase to form ATP.

The amount of mitochondria in a cell depends on how much energy that cell needs to produce. Muscle cells, for example, have many mitochondria because they need to produce energy to move the body. Red blood cells, which carry oxygen to other cells, have none; they do not need to produce energy. Mitochondria are analogous to a furnace or a powerhouse in the cell because, like furnaces and powerhouses, mitochondria produce energy from basic components (in this case, molecules that have been broken down so that they can be used).

Mitochondria have many other functions as well. They can store calcium, which maintains homeostasis of calcium levels in the cell. They also regulate the cell's metabolism and have roles in apoptosis (controlled cell death), cell signaling, and thermogenesis (heat production).

Structure of Mitochondria

Mitochondria have two membranes, an outer membrane and an inner membrane. These membranes are made of phospholipid layers, just like the cell's outer membrane. The outer membrane covers the surface of the mitochondrion, while the inner membrane is located within and has many folds called cristae. The folds increase surface area of the membrane, which is important because the inner membrane holds the proteins involved in the electron transport chain. It is also where many other chemical reactions take place to carry out the mitochondria's many functions. An increased surface area creates more space for more reactions to occur, and an increase the mitochondria's output. The space between the outer and inner membranes is called the intermembrane space, and the space inside the inner membrane is called the matrix.

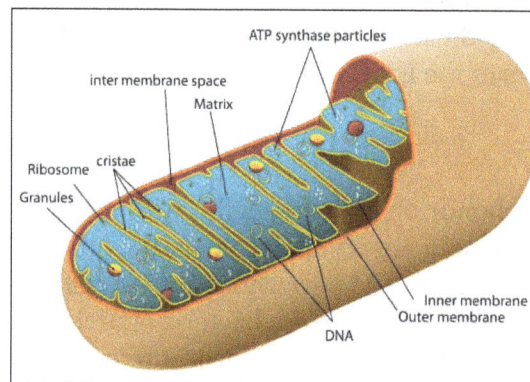

The Structure of a Mitochondrion

Cell Wall

The cell wall is an additional layer of protection on top of the cell membrane. You can find cell walls in both prokaryotes and eukaryotes, and they are most common in plants, algae, fungi and bacteria.

However, animals and protozoans do not have this type of structure. Cell walls tend to be rigid structures that help maintain the shape of the cell.

Function of a Cell Wall

The cell wall has several functions, including the maintenance of the cell structure and shape. The wall is rigid, so it protects the cell and its contents.

For example, the cell wall can keep pathogens like plant viruses from entering. In addition to the mechanical support, the wall acts as a framework that can prevent the cell from expanding or growing too quickly. Proteins, cellulose fibers, polysaccharides and other structural components help the wall maintain the shape of the cell.

The cell wall also plays an important role in transport. Since the wall is a semi-permeable membrane, it allows certain substances to pass through, such as proteins. This allows the wall to regulate diffusion in the cell and control what enters or leaves.

Additionally, the semi-permeable membrane helps communication among cells by allowing signaling molecules to pass through the pores.

Plant Cell Wall

A plant cell wall consists primarily of carbohydrates, like pectins, cellulose and hemicellulose. It also has structural proteins in smaller amounts and some minerals such as silicon. All of these components are vital parts of the cell wall.

Cellulose is a complex carbohydrate and consists of thousands of *glucose monomers* that form long chains. These chains come together and form cellulose *microfibrils*, which are several nanometers in diameter. The microfibrils help control the growth of the cell by limiting or allowing its expansion.

Turgor Pressure

One of the main reasons for having a wall in a plant cell is that it can withstand turgor pressure, and this is where cellulose plays a crucial role. Turgor pressure is a force created by the inside of the cell pushing out. Cellulose microfibrils form a matrix with the proteins, hemicelluloses and pectins to provide the strong framework that can resist turgor pressure.

Both hemicelluloses and pectins are branched polysaccharides. Hemicelluloses have hydrogen bonds connecting them to the cellulose microfibrils, while pectins trap water molecules to create a gel. Hemicelluloses increase the strength of the matrix and pectins help prevent compression.

Proteins in the Cell Wall

The proteins in the cell wall serve different functions. Some of them provide structural support. Others are enzymes, which are a type of protein that can speed up chemical reactions.

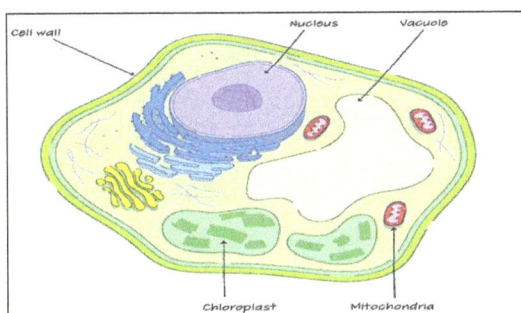

The enzymes help the formation of and normal modifications that occur to maintain the plant's cell wall. They also play a part in fruit ripening and leaf color changes.

If you have ever made your own jam or jelly, then you have seen the same types of pectins found in cell walls in action. Pectin is the ingredient that cooks add to thicken fruit juices. They often use the pectins naturally found in apples or berries to make their jams or jellies.

Structure of the Plant Cell Wall

Plant cell walls are three-layered structures with a *middle lamella*, *primary cell wall* and *secondary cell wall*. The middle lamella is the outermost layer and helps with cell-to-cell junctions while holding adjacent cells together (in other words, it sits between and holds together the cell walls of two cells; this is why it's called the middle lamella, even though it is the outermost layer).

The middle lamella acts like glue or cement for plant cells because it contains pectins. During cell division, the middle lamella is the first to form.

Primary Cell Wall

The primary cell wall develops when the cell grows, so it tends to be thin and flexible. It forms between the middle lamella and the *plasma membrane*.

It consists of cellulose microfibrils with hemicelluloses and pectins. This layer allows the cell to grow over time but does not overly restrict the cell's growth.

Secondary Cell Wall

The secondary cell wall is thicker and more rigid, so it provides more protection for the plant. It exists between the primary cell wall and the plasma membrane. Often, the primary cell wall actually helps create this secondary wall after the cell finishes growing.

Secondary cell walls consist of cellulose, hemicelluloses and *lignin*. Lignin is a polymer of aromatic alcohol that provides additional support for the plant. It helps protect the plant from attacks by insects or pathogens. Lignin also helps with water transport in the cells.

Difference between Primary and Secondary Cell Walls in Plants

When you compare the composition and thickness of primary and secondary cell walls in plants, it is easy to see the differences.

First, primary walls have equal amounts of cellulose, pectins and hemicelluloses. However, secondary cell walls do not have any pectin and have more cellulose. Second, the cellulose microfibrils in primary cells walls look random, but they are organized in secondary walls.

Although scientists have discovered many aspects of how cell walls function in plants, some areas still need more research.

For example, they are still learning more about the actual genes involved in the biosynthesis of the cell wall. Researchers estimate that about 2,000 genes take part in the process. Another important area of study is how gene regulation works in the plant cells and how it affects the wall.

Structure of Fungal and Algal Cell Walls

Similar to plants, the cell walls of fungi consist of carbohydrates. However, while fungi have cells with *chitin* and other carbohydrates, they do not have cellulose like plants do.

Their cell walls also have:

- Enzymes
- Glucans
- Pigments
- Waxes
- Other substances.

It is important to note that not all fungi have cell walls, but many of them do. In fungi, the cell wall sits outside the plasma membrane. Chitin makes up most of the cell wall, and it is the same material that gives insects their strong exoskeletons.

Fungal Cell Walls

In general, fungi with cell walls have three layers: chitin, glucans and proteins. As the innermost layer, chitin is fibrous and made up of polysaccharides. It helps make the fungi cell walls rigid and strong. Next, there is a layer of glucans, which are glucose polymers, crosslinking with chitin. The glucans also help the fungi maintain their cell wall rigidity. Finally, there is a layer of proteins called the mannoproteins or mannans, which have a high level of mannose sugar. The cell wall also has enzymes and structural proteins.

Different components of the fungal cell wall can serve different purposes. For instance, enzymes can help with digestion of organic materials, while other proteins may help with adhesion in the environment.

Cell Walls in Algae

The cell walls in algae consist of polysaccharides, like cellulose, or glycoproteins. Some algae have both polysaccharides and glycoproteins in their cell walls. In addition, algal cell walls have mannans, xylans, alginic acid and sulfonated polysaccharides. The cell walls among different types of algae can vary greatly.

Mannans are proteins that make microfibrils in some green and red algae. Xylans are complex polysaccharides and sometimes replace cellulose in algae. Alginic acid is another type of polysaccharide often found in brown algae. However, most algae have sulfonated polysaccharides.

Diatoms are a type of algae that live in water and soil. They are unique because their cell walls are made of silica. Researchers are still investigating how diatoms form their cell walls and which proteins make up the process. Nevertheless, they have determined that diatoms form their mineral-rich walls internally and move them outside the cell. This process, called *exocytosis*, is complex and involves multiple proteins.

Bacterial Cell Walls

A bacterial cell wall has peptidoglycans. Peptidoglycan or *murein* is a unique molecule that consists of sugars and amino acids in a mesh layer, and it helps the cell maintain its shape and structure.

The cell wall in bacteria exists outside of the plasma membrane. Not only does the wall help configure the shape of the cell, but it also helps prevent the cell from bursting and spilling all of its content.

Gram-positive and Gram-negative Bacteria

In general, you can divide bacteria into gram-positive or gram-negative categories, and each type has a slightly different cell wall. Gram-positive bacteria can stain blue or violet during a Gram staining test, which uses dyes to react with the peptidoglycans in the cell wall.

On the other hand, gram-negative bacteria cannot be stained blue or violet with this type of test. Today, microbiologists still use Gram staining to identify the type of bacteria. It is important to note that both gram-positive and gram-negative bacteria have peptidoglycans, but an extra outer membrane prevents the staining of gram-negative bacteria.

Gram-positive bacteria have thick cell walls made from layers of peptidoglycans. Gram-positive bacteria have one plasma membrane surrounded by this cell wall. However, gram-negative bacteria have thin cell walls of peptidoglycans that are not enough to protect them.

This is why gram-negative bacteria have an additional layer of *lipopolysaccharides* (LPS) that serve as an *endotoxin*. Gram-negative bacteria have an inner and outer plasma membrane, and the thin cell walls are in between the membranes.

Antibiotics and Bacteria

The differences between human and bacterial cells make it possible to use antibiotics in your body without killing all of your cells. Since people do not have cell walls, medications like antibiotics can target cell walls in bacteria. The composition of the cell wall plays a role in how some antibiotics work.

For example, penicillin, a common beta-lactam antibiotic, can affect the enzyme that forms the links between peptidoglycan strands in bacteria. This helps to destroy the protective cell wall and stops the bacteria from growing. Unfortunately, antibiotics can kill both helpful and harmful bacteria in the body.

Another group of antibiotics called glycopeptides targets the synthesis of cell walls by stopping peptidoglycans from forming. Examples of glycopeptide antibiotics include vancomycin and teicoplanin.

Antibiotic Resistance

Antibiotic resistance happens when bacteria change, which makes the drugs less effective. Since the resistant bacteria survive, they can reproduce and multiply. Bacteria become resistant to antibiotics in different ways.

For instance, they can change their cell walls. They can move the antibiotic out of their cells, or they can share genetic information that includes resistance to the drugs.

One way some bacteria resist beta-lactam antibiotics like penicillin is to make an enzyme called beta-lactamase. The enzyme attacks the beta-lactam ring, which is a core component of the drug, and consists of carbon, hydrogen, nitrogen and oxygen. However, drug manufacturers try to prevent this resistance by adding beta-lactamase inhibitors.

Cell Walls Matter

Cell walls offer protection, support and structural help for plants, algae, fungi and bacteria. Although there are major differences among the cell walls of prokaryotes and eukaryotes, most organisms have their cell walls outside of the plasma membranes.

Another similarity is that most cell walls provide rigidity and strength that help the cells maintain their shape. Protection from pathogens or predators is also something that many cell walls among different organisms have in common. Many organisms have cell walls made up of proteins and sugars.

Understanding the cell walls of prokaryotes and eukaryotes can help people in a variety of ways. From better medications to stronger crops, learning more about the cell wall offers a lot of potential benefits.

Cytoplasm

Cytoplasm refers to the fluid that fills the cell, which includes the cytosol along with filaments, proteins, ions and macromolecular structures as well as the organelles suspended in the cytosol.

In eukaryotic cells, cytoplasm refers to the contents of the cell with the exception of the nucleus. Eukaryotes have elaborate mechanisms for maintaining a distinct nuclear compartment separate from the cytoplasm. Active transport is involved in the creation of these subcellular structures and for maintaining homeostasis with the cytoplasm. For prokaryotic cells, since they do not have a defined nuclear membrane, the cytoplasm also contains the cell's primary genetic material. These cells are usually smaller in comparison to eukaryotes, and have a simpler internal organization of the cytoplasm.

Structure of Cytoplasm

The cytoplasm is unusual because it is unlike any other fluid found in the physical world. Liquids that are studied to understand diffusion usually contain a few solutes in an aqueous environment. However, the cytoplasm is a complex and crowded system containing a wide range of particles – from ions and small molecules, to proteins as well as giant multi protein complexes and organelles. These constituents are moved across the cell depending on the requirements of the cell along an elaborate cytoskeleton with the help of specialized motor proteins. The movement of such large particles also changes the physical properties of the cytosol.

The physical nature of the cytoplasm is variable. Sometimes, there is quick diffusion across the cell, making the cytoplasm resemble a colloidal solution. At other times, it appears to take on the properties of a gel-like or glass-like substance. It is said to have the properties of viscous as well as elastic materials – capable of deforming slowly under external force in addition to regaining its original shape with minimal loss of energy. Parts of the cytoplasm close to the plasma membrane are also 'stiffer' while the regions near the interior resemble free flowing liquids. These changes in the cytoplasm appear to be dependent on the metabolic processes within the cell and play an important role in carrying out specific functions and protecting the cell from stressors.

The cytoplasm can be divided into three components:

1. The cytoskeleton with its associated motor proteins.

2. Organelles and other large multi-protein complexes.

3. Cytoplasmic inclusions and dissolved solutes.

Cytoskeleton and Motor Proteins

The basic shape of the cell is provided by its cytoskeleton formed primarily by three types of polymers – actin filaments, microtubules and intermediate filaments.

Actin filaments or microfilaments are 7 nm in width and are made of double stranded polymers of F-actin. These filaments are associated with a number of other proteins that help in filament assembly and are also involved in anchoring them close to the plasma membrane. This cytoplasmic location helps the microfilaments become involved in rapid responses to signal molecules from the extracellular environment and produce cellular responses through signal transduction or chemotaxis. In addition, myosin, an ATP-based motor protein transmits cargo and vesicles along the microfilament and is also involved in muscle contraction.

Microtubules are polymers of α and β tubulin, which form a hollow tube by the lateral association of 13 protofilaments. Each protofilament is a polymer of alternating α and β tubulin molecules. The inner diameter of a microtubule is 12 nm and its outer diameter is 24 nm.

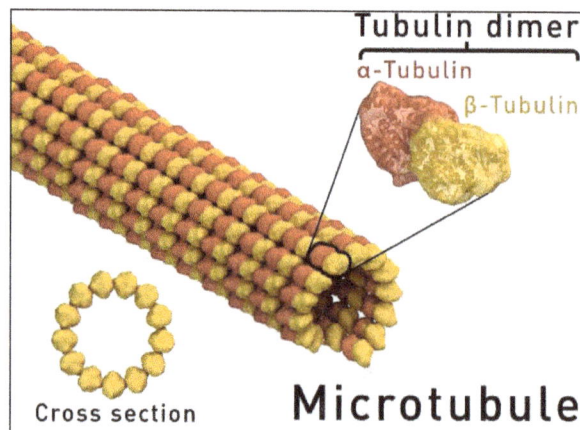

Microtubule structure

Microtubules radiate towards the periphery of the cell from microtubule organizing centers (MTOCs) located close to the nucleus, and provide structure and shape to the cell.

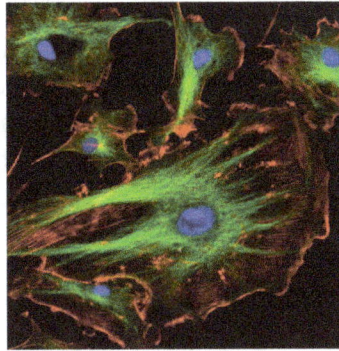

Fluorescent Cells

This figure shows the nucleus in blue, the actin filaments on the cell periphery are labeled red and the extensive microtubule network is marked green. The cytoplasm undergoes rapid reorganization during cell division with microtubules forming the spindle, which binds to chromosomes and segregates them into two daughter cells.

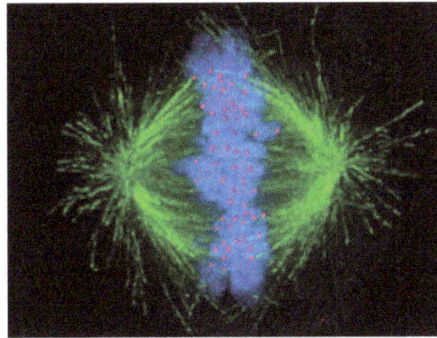

Kinetochore

Similar to the previous image, chromosomes are stained blue and microtubules are green. Tiny red dots are kinetochores.

Microtubules are involved in cytoplasmic transport, chromosome segregation and in forming structures such as cilia and flagella for cellular movement.

Intermediate filaments are larger than microfilaments but smaller than microtubules and are formed by a group of proteins that share structural features. Though they are not involved in cell motility, they are important for cells to come together as tissues and to remain anchored to the extracellular matrix.

Organelles and Multi-protein Complexes

Most eukaryotic cells have a number of organelles that provide compartments within the cytoplasm for specialized microenvironments. For instance, lysosomes contain a number of hydrolases in an acidic environment that is ideal for their enzymatic activity. These hydrolases are actively transported into the lysosome after being synthesized in the cytoplasm. Mitochondria, while containing their own genome, also need many enzymes synthesized in the cytosol, which are then selectively moved into the organelle. These organelles are placed in specific locations due to the physical gel-like nature of the cytoplasm and by anchoring to the cytoskeleton.

In addition, the cytoplasm also plays host to multi-protein complexes like the proteasome and ribosomes. Ribosomes are large complexes of RNA and protein that are important for the translation of mRNA code into amino acid sequences of proteins. Proteasomes are giant molecular structures about 20,000 kilodaltons in mass and 15 nm in diameter. Proteasomes are important for targeted destruction of proteins that are no longer needed by the cell.

Cytoplasmic Inclusions

Cytoplasmic inclusions can include a wide range of biochemicals – from small crystals of proteins, to pigments, carbohydrates and fats. All cells, especially in tissue like the adipose, contain droplets of lipids in their triglyceride form. These are used to create cellular membranes and are an excellent energy store. Lipids can generate twice as many ATP molecules per gram when compared to carbohydrates. However, the process of releasing this energy from triglycerides in intensive in oxygen consumption and therefore the cell also contains stores of glycogen as cytoplasmic inclusions. Glycogen inclusions are particularly important in cells like the skeletal and cardiac muscle cells where there can be a sudden increase in demand for glucose. Glycogen can be quickly broken down into individual molecules of glucose and used in cellular respiration before the cell can obtain more glucose reserves from the body.

Crystals are another type of cytoplasmic inclusion found in many cells, and have special function in cells of the inner ear (maintaining balance). Presence of crystals in cells of the testis appears to be linked with morbidity and infertility. Finally, the cytoplasm also contains pigments such as melanin, which lead to the pigmented cells of the skin. These pigments protect the cell and internal body structures from the deleterious effects of ultraviolet radiation. Pigments are also prominent in the cells of the iris that surround the pupil of the eye.

Each of these components affects the functioning of the cytoplasm in different ways, making it a dynamic region that plays a role in, and is influenced by, the cell's overall metabolic activity.

Functions of Cytoplasm

The cytoplasm is the site for most of the enzymatic reactions and metabolic activity of the cell. Cellular respiration begins in the cytoplasm with anaerobic respiration or glycolysis. This reaction provides the intermediates that are used by the mitochondria to generate ATP. In addition, the translation of mRNA into proteins on ribosomes also occurs mostly in the cytoplasm. Some of it happens on free ribosomes suspended in the cytosol while the rest happens on ribosomes anchored on the endoplasmic reticulum.

The cytoplasm also contains the monomers that go on to generate the cytoskeleton. The cytoskeleton, in addition to being important for the normal activities of the cell is crucial for cells that have a specialized shape. For instance, neurons with their long axons need the presence of intermediate filaments, microtubules, and actin filaments in order to provide a rigid framework for the action potential to be transmitted to the next cell. Additionally, some epithelial cells contain small cilia or flagella to move the cell or remove foreign particles through coordinated activity of cytoplasmic extrusions formed through the cytoskeleton.

The cytoplasm also plays a role in creating order within the cell with specific locations for different organelles. For instance, the nucleus is usually seen towards the center of the cell, with a centrosome nearby. The extensive endoplasmic reticulum and Golgi network are also placed in relation to the nucleus, with the vesicles radiating out towards the plasma membrane.

Cytoplasmic Streaming

Movement within the cytoplasm also occurs in bulk, through the directed movement of cytosol around the nucleus or vacuole. This is particularly important in large single celled organisms such as some species of green algae, which can be nearly 10 cm in length. Cytoplasmic streaming is also important for positioning chloroplasts close to the plasma membrane to ptimize photosynthesis and for distributing nutrients through the entire cell. In some cells, such as mouse oocytes, cytoplasmic streaming is expected to have a role in the formation of cellular sub-compartments and in organelle positioning as well.

Cytoplasmic Inheritance

The cytoplasm plays hosts to two organelles that contain their own genomes – the chloroplast and mitochondria. These organelles are inherited directly from the mother through the oocyte and therefore constitute genes that are inherited outside the nucleus. These organelles replicate independent of the nucleus and respond to the needs of the cell. Cytoplasmic or extra nuclear inheritance, therefore, forms an unbroken genetic line that has not undergone mixing or recombination with the male parent.

Cytoskeleton

The cytoskeleton is a structure that helps cells maintain their shape and internal organization, and it also provides mechanical support that enables cells to carry out essential functions like division and movement. There is no single cytoskeletal component. Rather, several different components work together to form the cytoskeleton.

Formation of Cytoskeleton

The cytoskeleton of eukaryotic cells is made of filamentous proteins, and it provides mechanical support to the cell and its cytoplasmic constituents. All cytoskeletons consist of three major classes of elements that differ in size and in protein composition. Microtubules are the largest type of filament, with a diameter of about 25 nanometers (nm), and they are composed of a protein called tubulin. Actin filaments are the smallest type, with a diameter of only about 6 nm, and they are made of a protein called actin. Intermediate filaments, as their name suggests, are mid-sized, with a diameter of about 10 nm. Unlike actin filaments and microtubules, intermediate filaments are constructed from a number of different subunit proteins.

Microtubules

Tubulin contains two polypeptide subunits, and dimers of these subunits string together to make long strands called protofilaments. Thirteen protofilaments then come together to form the hollow, straw-shaped filaments of microtubules. Microtubules are ever-changing, with reactions constantly adding and subtracting tubulin dimers at both ends of the filament. The rates of change at either end are not balanced — one end grows more rapidly and is called the plus end, whereas the other end is known as the minus end. In cells, the minus ends of microtubules are anchored in structures called microtubule organizing centers (MTOCs). The primary MTOC in a cell is called the centrosome, and it is usually located adjacent to the nucleus.

Microtubules tend to grow out from the centrosome to the plasma membrane. In nondividing cells, microtubule networks radiate out from the centrosome to provide the basic organization of the cytoplasm, including the positioning of organelles.

Actin Filaments

The protein actin is abundant in all eukaryotic cells. It was first discovered in skeletal muscle, where actin filaments slide along filaments of another protein called myosin to make the cells contract. (In nonmuscle cells, actin filaments are less organized and myosin is much less prominent.) Actin filaments are made up of identical actin proteins arranged in a long spiral chain. Like microtubules, actin filaments have plus and minus ends, with more ATP-powered growth occurring at a filament's plus end.

In many types of cells, networks of actin filaments are found beneath the cell cortex, which is the meshwork of membrane-associated proteins that supports and strengthens the plasma membrane. Such networks allow cells to hold — and move — specialized shapes, such as the brush border of microvilli. Actin filaments are also involved in cytokinesis and cell movement.

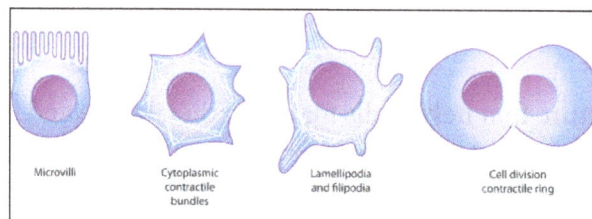

Actin filaments support a variety of structures in a cell

Intermediate Filaments

Intermediate filaments come in several types, but they are generally strong and ropelike. Their functions are primarily mechanical and, as a class, intermediate filaments are less dynamic than actin filaments or microtubules. Intermediate filaments commonly work in tandem with microtubules, providing strength and support for the fragile tubulin structures.

All cells have intermediate filaments, but the protein subunits of these structures vary. Some cells have multiple types of intermediate filaments, and some intermediate filaments are associated with specific cell types. For example, neurofilaments are found specifically in neurons (most prominently in the long axons of these cells), desmin filaments are found specifically in muscle cells, and keratins are found specifically in epithelial cells. Other intermediate filaments are distributed more widely. For example, vimentin filaments are found in a broad range of cell types and frequently colocalize with microtubules. Similarly, lamins are found in all cell types, where they form a meshwork that reinforces the inside of the nuclear membrane. Note that intermediate filaments are not polar in the way that actin or tubulins are.

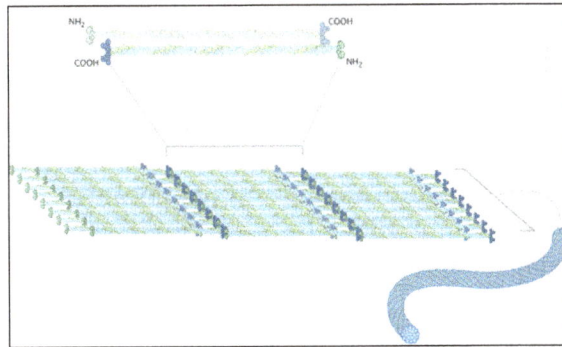

The structure of intermediate filaments: Intermediate filaments are composed of smaller strands in the shape of rods. Eight rods are aligned in a staggered array with another eight rods, and these components all twist together to form the rope-like conformation of an intermediate filament.

Cell Movement

Cytoskeletal filaments provide the basis for cell movement. For instance, cilia and (eukaryotic) flagella move as a result of microtubules sliding along each other. In fact, cross sections of these tail-like cellular extensions show organized arrays of microtubules.

Other cell movements, such as the pinching off of the cell membrane in the final step of cell division (also known as cytokinesis) are produced by the contractile capacity of actin filament networks. Actin filaments are extremely dynamic and can rapidly form and disassemble. In fact, this dynamic action underlies the crawling behavior of cells such as amoebae. At the leading edge of a moving cell, actin filaments are rapidly polymerizing; at its rear edge, they are quickly depolymerizing. A large number of other proteins participate in actin assembly and disassembly as well.

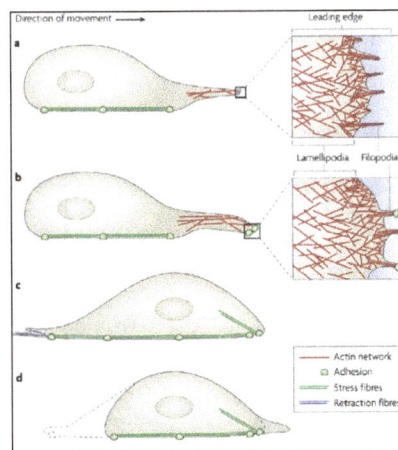

Cell migration is dependent on different actin filament structures

In figure, (A) In a cell, motility is initiated by an actin-dependent protrusion of the cell's leading edge, which is composed of armlike structures called lamellipodia and filopodia. These protrusive structures contain actin filaments, with elongating barbed ends orientated toward the plasma membrane. (B) During cellular arm extension, the plasma membrane sticks to the surface at the leading edge. (C) Next, the nucleus and the cell body are pushed forward through intracellular contraction forces mediated by stress fibers. (D) Then, retraction fibers pull the rear of the cell forward.

Cell Junctions

Cell junctions can be divided into two types: those that link cells together, also called intercellular junctions (tight, gap, adherens, and desmosomal junctions), and those that link cells to the extracellular matrix (focal contacts/adhesion plaques and hemidesmosomes). These junctions play a prominent role in maintaining the integrity of tissues in multicellular organisms and some, if not all of them, are involved in signal transduction.

Intercellular junctions and hemidesmosomes were first identified in tissues examined by electron microscopy. In contrast, the focal contact was first observed in cultured cells in the light microscope by a technique called interference reflection. This procedure revealed specific sites where cells closely adhere to their substrate. These were called focal contacts oradhesion plaques.

Tight Junctions

The tight junction (also referred to as a zonula occludens) is a site where the membranes of two cells come very close together. In fact, the outer leaflets of the membranes of the contacting cells appear to be fused. Tight junctions, as their name implies, act as a barrier so that materials cannot pass between two interacting cells. The protein components of the tight junction are arranged like beads on a string that span the adjacent membranes of each tight junction.

Tight junctions often occur in a belt completely encircling the cell. In a sheet of such cells, material cannot pass from one side of the sheet to the other by squeezing between cells. Instead, it must go through a cell, and hence the cell can regulate its passage. Such an arrangement is found in the gut, to regulate absorption of digested nutrients.

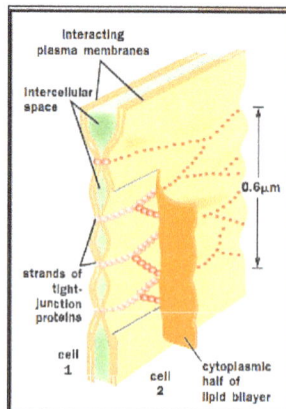

A model of a tight junction. It is thought that the strands that hold adjacent plasma membranes together are formed by continuous strands of transmembrane junctional proteins across the intercellular space, creating a seal.

Gap Junctions

In contrast to the tight junction, there is a channel between the membranes of contacting cells in the gap junction so that the cytoplasm of the two is connected. The basic building block of each gap junction is the connexin subunit. Six of these in each of the membranes of two neighboring cells come together, and then the group of six connexins in one cell interact with a omparable hexamer in the other cell resulting in the formation of a channel. This channel allows direct cytoplasmic communication among the cells; small molecules of 1,500 daltons or less can pass through the channel of each gap junction whose opening or closing can be controlled locally in the cell. Gap junctions unite muscle cells in the heart to help coordinate their contraction.

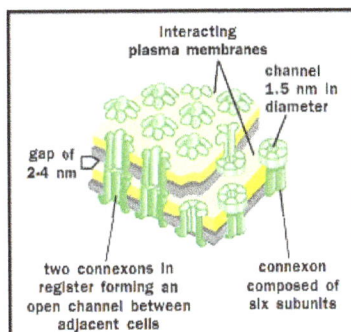

Model of a gap junction. The lipid bilayers are penetrated by protein assemblies called connexons. Two connexons join across the intercellular space to form a continuous aqueous channel that links the two cells.

Adherens Junctions and Focal Contacts

Adherens junctions (sometimes called zonula adherens) are found at sites of cell-cell interaction. Focal contacts mediate association of cells with the extracellular matrix. Both associate with the actin cytoskeleton and both are involved in adhesion (sticking cells together or sticking cells to surfaces). Focal contacts possess specific transmembrane receptors of the integrin family that link the cell to the extracellular matrix on the outside of the cell and the microfilament system on the inside. Conversely, members of a family of calcium ion-dependent cell adhesion molecules, called cadherins, mediate attachment between cells at adherens junctions. Adherens junctions and focal contacts not only tether cells together or to the extracellular matrix, but they also transduce signals into and out of the cell, influencing a variety of cellular behaviors including proliferation, migration, and differentiation. In fact some protein components of these junctions can shuttle to and from the nucleus where they are thought to play a role in regulating gene expression.

Desmosomes and Hemidesmosomes

Desmosomes (the macula adherens) and hemidesmosomes are distinguished by their association with the keratin -based cytoskeleton. Despite their names, desmosomes and hemidesmosomes are distinct at the molecular level. Both are primarily involved in adhesion. The desmosome, like the adherens junction, possesses calcium ion-dependent cell adhesion molecules that interact with similar molecules in the adjacent cell. Meanwhile, integrins at the core of the hemidesmosome mediate its interaction with the extracellular matrix. The hemidesmosome and, most likely, the desmosome are also sites of signal transduction

Centrioles

Centrioles occur as paired cylindrical organelles together with pericentriolar material (PCM) in the centrosome of an animal cell. Centrioles are found as single structures in cilia and flagella in animal cells and some lower plant cells.

Centrioles are present in (1) animal cells and (2) the basal region of cilia and flagella in animals and lower plants (e.g. chlamydomonas). In cilia and flagella centrioles are called 'basal bodies' but the two can be considered inter-convertible.

Centrioles are absent from the cells of higher plants.

When animal cells undergo mitosis they are considered by some to benefit from the presence of centrioles which appear to control spindle fibre formation and which later has an effect on chromosome separation. Research however has shown that mitosis can take place in animal cells after centrioles have been destroyed. Sometimes this seems to be at the expense of abnormalities in spindle development and subsequent problems with chromosome separation. Recent research also suggests that embryos of *Drosophila* arrest very early if centriole replication cannot take place.

In higher plants mitosis takes place perfectly satisfactorily with microtubules forming spindle fibres but without the help of centrioles. The function of centrioles therefore remains something of a mystery.

Structure

A centriole is composed of short lengths of microtubules arranged in the form of an open-ended cylinder about 500nm long and 200nm in diameter. The microtubules forming the wall of the cylinder are grouped into nine sets of bundles of three microtubules each.

In cilia and flagella where centrioles are at the base of the structure, and are called basal bodies, the wall and cavity architecture is slightly different. In addition to cylinder walls composed of nine sets of bundles of three microtubules, there are walls of nine sets of two bundles. In both types there is a central matrix from which spokes radiate as in a cart wheel.

In animal cells centrioles usually reside in pairs with the cylindrical centrioles at right angles to each other.

Centrioles organise a 'cloud' of protein material around themselves; this is the pericentriolar material (PCM). Together the two constitute the all-important centrosome.

Function

Centrioles function as a pair in most cells in animals but as a single centriole or basal body in cilia and flagella.

Centrioles in Pairs

Cells entering mitosis have a centrosome containing two pairs of centrioles and associated pericentriolar material (PCM). During prophase the centrosome divides into two parts and a centriole pair migrates to each end or pole on the outside of the nuclear membrane or envelope. At this point microtubules are produced at the outer edge of the pericentriolar material and grow out in a radial form. The centriole pair and PCM is called an aster. Microtubules from the aster at one pole grow towards the aster at the opposite pole. These microtubules are called spindle fibres. Some of these will become attached by centromeres to chromosomes lined up on the 'equator' of the dividing cell. Others, though not attached to chromatids/chromosomes by centromeres, will assist in pushing apart the two parts of the dividing cell.

A Single Centriole or Basal Body

At the base of each cilium or flagellum there is a single centriole. This structure and associated pericentriolar material, construct microtubules in a linear direction. These microtubules form most of the inside of cilia and flagella and are largely responsible, using protein motors, for the mechanical aspects of their movement. The centriole at the base of each one also appears to exert some degree of direction and control over the movement of the cilia and flagella.

Replication

In cells where centrioles are present as a pair, replication takes place during the whole of the cell cycle. In phase G1 the two centriole cylinders move very slightly apart from one another. During S phase new cylinders of microtubules form near, and at right angles to, the two 'mother' cylinders. The two pairs of centrioles keep very close to one another until the prophase stage of mitosis. At this point they separate with both pairs of centrioles moving over the outer surface of the nuclear envelope to opposite ends or 'poles' of the cell, to form the astral poles of the dividing cell.

Cytosol

Cytosol is the fluid that's found inside living cells. More specifically, it's the water-based fluid in which organelles, proteins and other structures of the cell live. Also known as the cytoplasmic matrix, it constitutes most of the intracellular fluid (ICF). Cytosol is often confused with cytoplasm, however, which is an entirely different entity within a cell.

While cytosol refers to the water and anything soluble that is dissolved in it, such as soluble proteins and ions, the cytoplasm consists of cytosol and certain insoluble suspended particles (e.g., ribosomes).

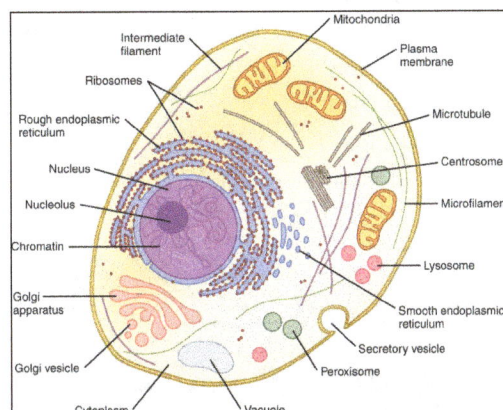

Cell and its organelles

As you may already know, the cell is the basic unit of life. Every living being in the world is made of billions upon billions of cells, which are responsible for a number of essential life processes. The function of these cells is determined by tiny organs (or organelles) that are present inside the cell.

Cytosol is a crucial cellular component that provides structural support for these organelles. A cell is like a plastic bag filled with fluid. The fluid it contains is the cytosol, while the bag itself is the cell membrane.

Cytosol Structure

The majority of cytosol is water, which makes up nearly 70% of the total volume of a cell. While the pH of the intracellular fluid is 7.4, human cytosolic pH lies between 7.0-7.4, and is typically higher when the cell is growing. In addition to water, cytosol also consists of small molecules, dissolved ions and large water-soluble water molecules (e.g., proteins).

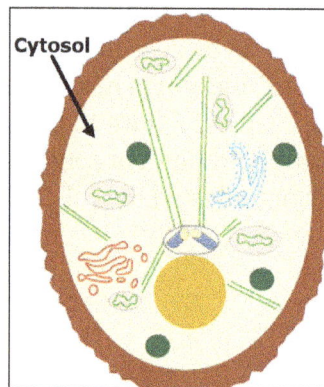

Cytosol consists of dissolved ions and water-soluble molecules

Note that the concentrations of the other ions present in the cytosol are different from those ions present in the extracellular fluid. In addition to that, higher amounts of charged macromolecules (e.g., proteins) and nucleic acids are present inside the cytosol than outside of the cell structure.

Cytosol Functions

Interestingly, cytosol does not have a single, well-defined function; rather, it serves as a site for a number of intracellular processes (e.g., signal transduction – a cellular mechanism that converts a stimulus into a response within the cell, from the cell membrane to sites present within the cell).

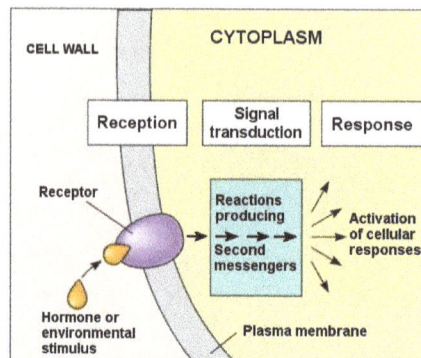

Signal transduction pathways model

A number of enzymatic activities require cytosol, as enzymes often require certain pH levels, salt concentrations and other environmental conditions that are appropriately fulfilled by cytosol. In addition to that, cytosol provides structural support to organelles. In fact, most cells depend on the volume of cytosol to make space for chemicals to move within the cell.

Cytosol vs. Cytoplasm

Cytosol is often confused with cytoplasm, but these two are entirely different entities related to a living cell. While cytoplasm consists of all the contents found inside a cell (excluding the nucleus), cytosol is just the liquid or aqueous part of the cytoplasm. In other words, cytoplasm is the area of space outside the nucleus and consists of cytosol and other organelles.

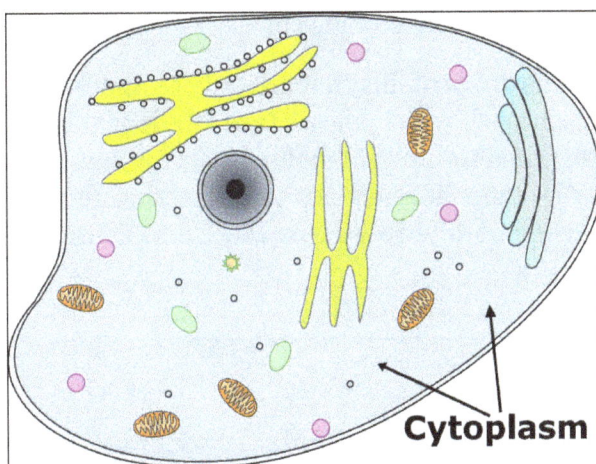

Cytoplasm of a cell

It is important to note that cytosol is a critical element of the cytoplasm. In a prokaryotic cell, cytosol is the host of almost all chemical reactions and metabolic processes that take place within the cell. Also, cytosol is the site for cell communication, while cytoplasm plays host to certain large-scale processes, such as cell division and glycolysis.

Membrane Proteins

The plasma membrane contains molecules other than phospholipids, primarily other lipids and proteins. The green molecules in figure below, for example, are the lipid cholesterol. Molecules of cholesterol help the plasma membrane keep its shape. Many of the proteins in the plasma membrane assist other substances in crossing the membrane.

The plasma membranes also contain certain types of proteins. A membrane protein is a protein molecule that is attached to, or associated with, the membrane of a cell or an organelle. Membrane proteins can be put into two groups based on how the protein is associated with the membrane.

Integral membrane proteins are permanently embedded within the plasma membrane. They have a range of important functions. Such functions include channeling or transporting molecules across the membrane. Other integral proteins act as cell receptors. Integral membrane proteins can be classified according to their relationship with the bilayer:

- Transmembrane proteins span the entire plasma membrane. Transmembrane proteins are found in all types of biological membranes.

- Integral monotopic proteins are permanently attached to the membrane from only one side.

Some integral membrane proteins are responsible for cell adhesion (sticking of a cell to another cell or surface). On the outside of cell membranes and attached to some of the proteins are carbohydrate chains that act as labels that identify the cell type. Shown in figure below are two different types of membrane proteins and associated molecules.

Peripheral membrane proteins are proteins that are only temporarily associated with the membrane. They can be easily removed, which allows them to be involved in cell signaling. Peripheral proteins can also be attached to integral membrane proteins, or they can stick into a small portion of the lipid bilayer by themselves. Peripheral membrane proteins are often associated with ion channels and transmembrane receptors. Most peripheral membrane proteins are hydrophilic.

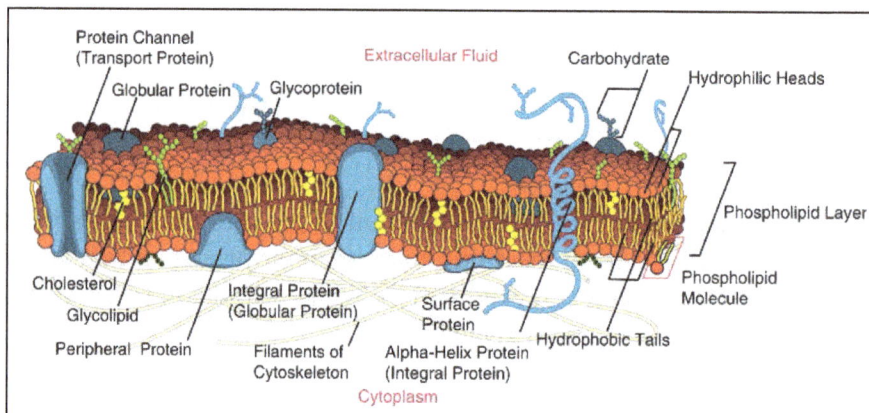

Some of the membrane proteins make up a major transport system that moves molecules and ions through the polar phospholipid bilayer.

Fluid Mosaic Model

In 1972 S.J. Singer and G.L. Nicolson proposed the now widely accepted Fluid Mosaic Modelof the structure of cell membranes. The model proposes that integral membrane proteins are embedded in the phospholipid bilayer, as seen in figure above. Some of these proteins extend all the way through the bilayer and some only partially across it. These membrane proteins act as transport proteins and receptors proteins.

Their model also proposed that the membrane behaves like a fluid, rather than a solid. The proteins and lipids of the membrane move around the membrane, much like buoys in water. Such movement causes a constant change in the "mosaic pattern" of the plasma membrane.

Extensions of the Plasma Membrane

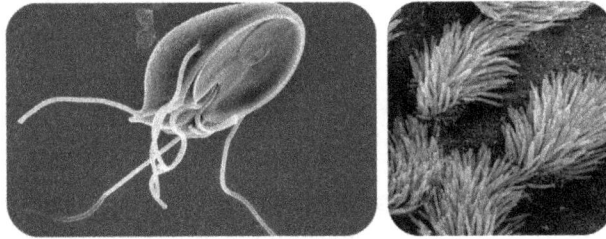

Flagella and Cilia. Cilia and flagella are extensions of the plasma membrane of many cells

The plasma membrane may have extensions, such as whip-like flagella or brush-like cilia. In single-celled organisms, like those shown in figure below, the membrane extensions may help the organisms move. In multicellular organisms, the extensions have other functions. For example, the cilia on human lung cells sweep foreign particles and mucus toward the mouth and nose.

References

- Cells-the-basic-units-of-life, lifesciences, science, read: siyavula.com, Retrieved 31 July, 2019

- Cell-organelle, the-fundamental-unit-of-life, biology, guides: toppr.com, Retrieved 25 August, 2019

- Structure-and-function-of-the-cell-nucleus, life-sciences: news-medical.net, Retrieved 20 July, 2019

- Ribosome, softcell-e-learning: bscb.org, Retrieved 23 June, 2019

- Golgi-apparatus: microscopemaster.com, Retrieved 15 May, 2019

- Endoplasmic-reticulum, science: britannica.com, Retrieved 12 April, 2019

- Mitochondria: biologydictionary.net, Retrieved 21 March, 2019

- Cell-wall-definition-structure-function-with-diagram: sciencing.com, Retrieved 13 July, 2019

- Cytoplasm: biologydictionary.net, Retrieved 16 June, 2019

- Microtubules-and-filaments, topicpage, scitable: nature.com, Retrieved 24 March, 2019

- Cell-junctions, ce-co: biologyreference.com, Retrieved 17 February, 2019

- Centriole, softcell-e-learning, learning-resources: bscb.org, Retrieved 21 April, 2019

- What-is-cytosol-how-is-it-different-from-cytoplasm, pure-sciences: scienceabc.com, Retrieved 13 July, 2019

Chapter 4

Cell Signaling and Cell Communication

The communication process which governs basic activities of cells and coordinates multiple cell actions is known as cell signaling. They are classified as either biochemical or mechanical in nature, based on the kind of signal. This chapter discusses in detail the processes and concepts related to cell signaling and communication.

Cell communication is the process by which a cell detects and responds to signals in its environment. Most single-celled organisms can perceive changes in nutrient availability and adapt their metabolism as needed. Some single-celled organisms may utilize environmental signals to locate a suitable mate; some send signals to make their numbers known to other members of their species.

The cells of multi-celled organisms must communicate with one another to coordinate the activities of the organism as a whole. Most multi-celled organisms can utilize junctions between cells for direct intercellular signaling. But many forms of communication entail binding of a signal molecule to the receptors of target cells. Receptor-mediated signaling can be short-range (affecting only nearby cells) or long-range (affecting cells throughout the organism).

Mechanism of Cell Signaling

In all cell signaling systems, the signaling molecule must bind to a specific receptor; this activates a signal transduction pathway which produces the cellular response. In the majority of signaling systems, the receptor is located within the plasma membrane. Enzyme-linked receptors are very common and occur to some extent in all known species. G-protein coupled receptors are common in eukaryotic cells. The ligand-gated channels that are used in membrane transport may also serve as receptors.

Transduction pathways vary in length and are designed to refine and/or amplify the signal. In bacterial cells, two-component regulatory systems are a common form of transduction pathway. In eukaryotic cells, the transduction pathways are usually longer and may include second messengers; two common second messengers are cAMP and calcium ions. Signal transduction pathways allow cells to respond differently to the same signaling molecule. In another type of signaling system, the receptors are located within the cytosol or nucleus of the cell; this results in the transcription of a gene.

Use of Hormones for Communication by Animals

Long-range communication through use of hormones is one of the two major methods of cell communication in animals; this is the function of the endocrine system. There are three classes of hormones which differ in their chemical structure and water solubility. These hormones also differ in their mechanism of action: protein/peptide hormones and most amines act by binding to the plasma membrane, whereas steroid hormones and thyroxine act within the nucleus.

Some hormones have important metabolic functions and are linked to physiological disorders. Thyroid hormones control metabolic rate; their levels are regulated by feedback loops. Insulin and glucagon control blood glucose levels; a malfunctioning pancreas can cause diabetes, and obesity often leads to another form of this disease.

Cell Signaling

Cell signaling can occur through a number of different pathways, but the overall theme is that the actions of one cell influence the function of another. Cell signaling is needed by multicellular organisms to coordinate a wide variety of functions. Nerve cells must communicate with muscle cells to create movement, immune cells must avoid destroying cells of the body, and cells must organize during the development of a baby.

Some forms of cell signaling are intracellular, while others are intercellular. Intracellular signals are produced by the same cell that receives the signal. On the other hand, intercellular signals can travel all throughout the body. This allows certain glands within the body to produce signals which take action on many different tissues across the body. Cell signaling is how a tiny gland within the brain can react to external stimuli and coordinate a response. In response to stimuli like light, odors, or touch, the gland can, in turn, release a hormone which activates responses in diverse body systems to coordinate a response to a threat or opportunity.

Cell Signaling Function and Key Players

Cell signaling serves a vital purpose in allowing our cells to carry out life as we know it. Moreover, thanks to the concerted efforts of our cells via their signaling molecules, our body is able to orchestrate the many complexities that maintain life. These complexities, in effect, demand a diverse collection of receptor-mediated pathways that execute their unique functions. In general, a ligand will activate a receptor and cause a specific response.

Intracellular Receptors

A common type of signaling receptor is the intracellular receptor, which is located within the cytoplasm of the cell and generally includes two types. In addition to cytoplasmic receptors, nuclear receptors are a special class of protein with diverse DNA binding domainthat when bound to steroid or thyroid hormones form a complex that enters the nucleus and modulates the transcription of a gene. IP_3 receptors are another class, which are located in the endoplasmic reticulum and carry out important functions like the release of Ca^{2+} that is so crucial for the contraction of our muscles and plasticity of our neural cells.

Ligand-gated Ion Channels

Spanning our plasma membranes are another type of receptor called Ligand-gated ion channels that allow hydrophilic ions to cross the thick fatty membranes of our cellsand organelles. When bound to a neurotransmitter like acetylcholine, ions (commonly K^+, Na^+, Ca^{2+}, or Cl^-) are allowed to flow through the membrane to allow the life-sustaining function of neural firing to take place, among many other functions.

G-protein Coupled Receptors

Comparatively, G-protein coupled receptors (GPCRs) remain the largest and most diverse group of membrane receptors in eukaryotes. In fact, they are special in that they receive input from a diverse group of signals ranging from light energy to peptides and sugars. In effect, their mechanism of action also starts with a ligand binding to its receptor. However, the demarcation is that ligand binding results in the activation of a G protein that is then able to transmit an entire cascade of enzyme and second messenger activations that carry out an incredible array of functions like sight, sensation, inflammation, and growth.

Receptor Tyrosine Kinases

Likewise, receptor tyrosine kinases (RTKs) are another class of receptors revealed to show diversity in their actions and mechanisms of activation. For example, the general method of activation follows a ligand binding to the receptor tyrosine kinase, which allows their kinase domains to dimerize. Then, this dimerization invites the phosphorylation of their tyrosine kinase domains that, in turn, allow intracellular proteins to bind the phosphorylated sites and become "active." An important function of receptor tyrosine kinases is their roles in mediating growth pathways. Of course, the downside of having complex signaling networks lies in the unforeseen ways in which any alteration can produce disease or unregulated growth – cancer. Still, much is yet to be understood about cell signaling pathways, but one appreciable fact is that the importance they carry is nothing short of monumental.

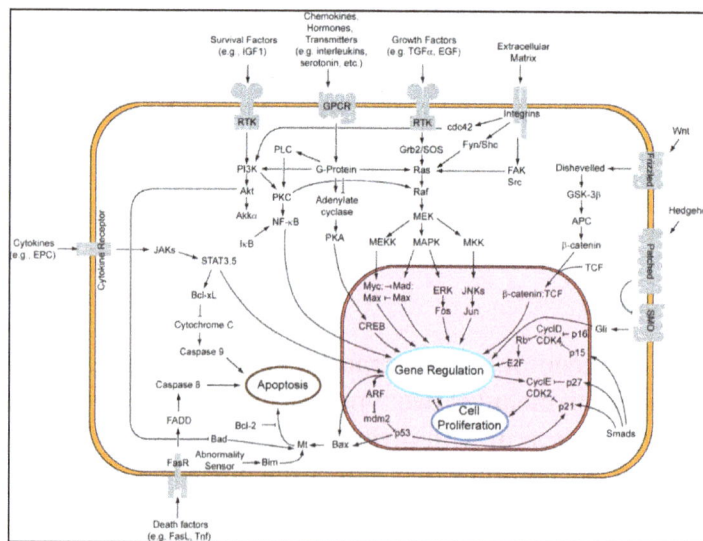

Signal transduction

For example, this figure depicts various forms of RTK and GPCR-mediated Signal Transduction.

Cell Signaling Pathways

Typically, cell signaling is either mechanical or biochemical and can occur locally. Additionally, categories of cell signaling are determined by the distance a ligand must travel. Likewise, hydrophobic ligands have fatty properties and include steroid hormones and vitamin D_3. These molecules are able to diffuse across the target cell's plasma membrane to bind intracellular receptors inside. On the other hand, hydrophilic ligands are often amino-acid derived. Instead, these molecules will bind to receptors on the surface of the cell. Comparatively, these polar molecules allow the signal to travel through the aqueous environment of our bodies without assistance.

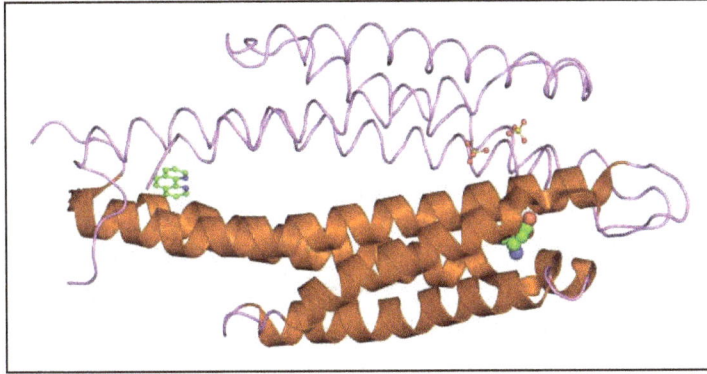

Protein molecular structure: The image depicts the molecular structure of a signal molecule binding to a target protein.

Types of Cell Signaling Molecules

Signaling molecules are currently assigned one of five classifications:

1. Intracrine ligands are produced by the target cell. Then, they bind to a receptor within the cell.

2. Autocrine ligands are distinct in that they function internally and on other target cells (ex. Immune cells).

3. Juxtacrine ligands target adjacent cells (often called "contact-dependent" signaling).

4. Paracrine ligands target cells only in the vicinity of the original emitting cell (ex. Neurotransmitters).

5. Lastly, Endocrine cells produce hormones that have the important task of targeting distant cells and often travel through our circulatory system.

Stages of Cell Signaling

Cell signaling can be divided into 3 stages:

1. Reception: A cell detects a signaling molecule from the outside of the cell. A signal is detected when the chemical signal (also known as a ligand) binds to a receptor protein on the surface of the cell or inside the cell.

2. Transduction: When the signaling molecule binds the receptor it changes the receptor protein in some way. This change initiates the process of transduction. Signal transduction is usually a pathway of several steps. Each relay molecule in the signal transduction pathway changes the next molecule in the pathway.

3. Response: Finally, the signal triggers a specific cellular response.

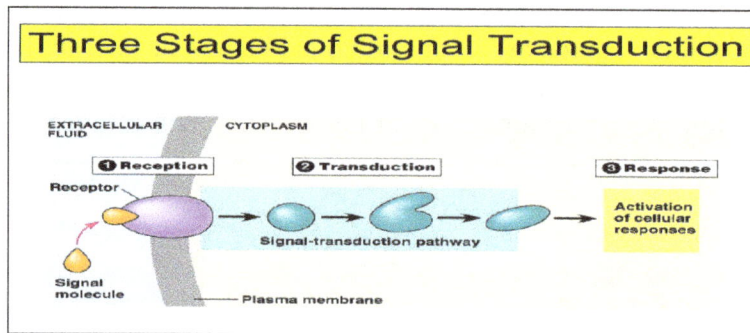

Three Stages of Cell Signaling

Membrane receptors function by binding the signal molecule (ligand) and causing the production of a second signal (also known as a second messenger) that then causes a cellular response. These type of receptors transmit information from the extracellular environment to the inside of the cell by changing shape or by joining with another protein once a specific ligand binds to it. Examples of membrane receptors include G Protein-Coupled Receptors and Receptor Tyrosine Kinases.

Intracellular receptors are found inside the cell, either in the cytopolasm or in the nucleus of the target cell (the cell receiving the signal). Chemical messengers that are hydrophobic or very small (steroid hormones for example) can pass through the plasma membrane without assistance and bind these intracellular receptors. Once bound and activated by the signal molecule, the activated receptor can initiate a cellular response, such as a change in gene expression.

Transduction

Since signaling systems need to be responsive to small concentrations of chemical signals and act quickly, cells often use a multi-step pathway that transmits the signal quickly, while amplifying the signal to numerous molecules at each step.

Steps in the signal transduction pathway often involve the addition or removal of phosphate groups which results in the activation of proteins. Enzymes that transfer phosphate groups from ATP to a protein are called protein kinases. Many of the relay molecules in a signal transduction pathway are protein kinases and often act on other protein kinases in the pathway. Often this creates a phosphorylation cascade, where one enzyme phosphorylates another, which then phosphorylates another protein, causing a chain reaction.

Also important to the phosphorylation cascade are a group of proteins known as protein phosphatases. Protein phosphatases are enzymes that can rapidly remove phosphate groups from proteins (dephosphorylation) and thus inactivate protein kinases. Protein phosphatases are the "off switch" in the signal transduction pathway. Turning the signal transduction pathway off when the signal is no longer present is important to ensure that the cellular response is regulated appropriately. Dephosphorylation also makes protein kinases available for reuse and enables the cell to respond again when another signal is received.

Kinases are not the only tools used by cells in signal transduction. Small, nonprotein, water-soluble molecules or ions called second messengers (the ligand that binds the receptor is the first messenger) can also relay signals received by receptors on the cell surface to target molecules in the cytoplasm or the nucleus. Examples of second messengers include cyclic AMP (cAMP) and calcium ions.

Response

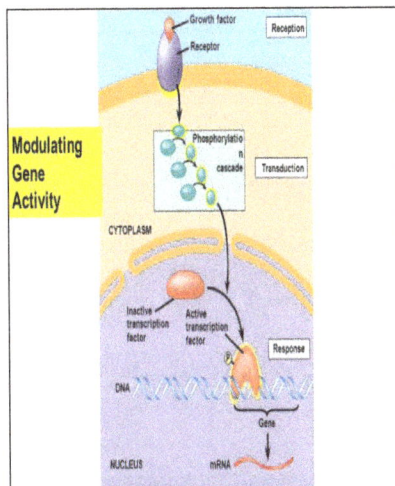

Modulating Gene Activity

Cell signaling ultimately leads to the regulation of one or more cellular activities. Regulation of gene expression (turning transcription of specific genes on or off) is a common outcome of cell signaling. A signaling pathway may also regulate the activity of a protein, for example opening or closing an ion channel in the plasma membrane or promoting a change in cell metabolism such as catalyzing the breakdown of glycogen. Signaling pathways can also lead to important cellular events such as cell division or apoptosis (programmed cell death).

Signaling Molecules and Cellular Receptors

In order to properly respond to external stimuli, cells have developed complex mechanisms of communication that can receive a message, transfer the information across the plasma membrane, and then produce changes within the cell in response to the message.

In multicellular organisms, cells send and receive chemical messages constantly to coordinate the actions of distant organs, tissues, and cells. The ability to send messages quickly and efficiently enables cells to coordinate and fine-tune their functions.

While the necessity for cellular communication in larger organisms seems obvious, even single-celled organisms communicate with each other. Yeast cells signal each other to aid mating. Some forms of bacteria coordinate their actions in order to form large complexes called biofilms or to organize the production of toxins to remove competing organisms. The ability of cells to communicate through chemical signals originated in single cells and was essential for the evolution of multicellular organisms. The efficient and error-free function of communication systems is vital for all forms of life.

Forms of Signaling

There are four categories of chemical signaling found in multicellular organisms: paracrine signaling, endocrine signaling, autocrine signaling, and direct signaling across gap junctions. The main difference between the different categories of signaling is the distance that the signal travels through the organism to reach the target cell. It is also important to note that not all cells are affected by the same signals.

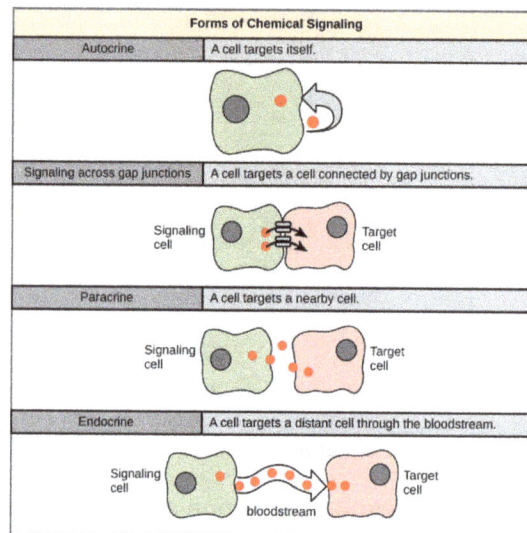

Forms of Chemical Signaling

In figure, in chemical signaling, a cell may target itself (autocrine signaling), a cell connected by gap junctions, a nearby cell (paracrine signaling), or a distant cell (endocrine signaling). Paracrine signaling acts on nearby cells, endocrine signaling uses the circulatory system to transport ligands, and autocrine signaling acts on the signaling cell. Signaling via gap junctions involves signaling molecules moving directly between adjacent cells.

Paracrine Signaling

Signals that act locally between cells that are close together are called paracrine signals. Paracrine signals move by diffusion through the extracellular matrix. These types of signals usually elicit quick responses that last only a short amount of time. In order to keep the response localized, paracrine ligand molecules are normally quickly degraded by enzymes or removed by neighboring cells. Removing the signals will reestablish the concentration gradient for the signal, allowing them to quickly diffuse through the intracellular space if released again.

One example of paracrine signaling is the transfer of signals across synapses between nerve cells. A nerve cell consists of a cell body, several short, branched extensions called dendrites that receive stimuli, and a long extension called an axon, which transmits signals to other nerve cells or muscle cells. The junction between nerve cells where signal transmission occurs is called a synapse. A synaptic signal is a chemical signal that travels between nerve cells. Signals within the nerve cells are propagated by fast-moving electrical impulses. When these impulses reach the end of the axon, the signal continues on to a dendrite of the next cell by the release of chemical ligands called neurotransmitters by the presynaptic cell (the cell emitting the signal). The neurotransmitters are transported across the very small distances between nerve cells, which are called chemical synapses. The small distance between nerve cells allows the signal to travel quickly; this enables an immediate response.

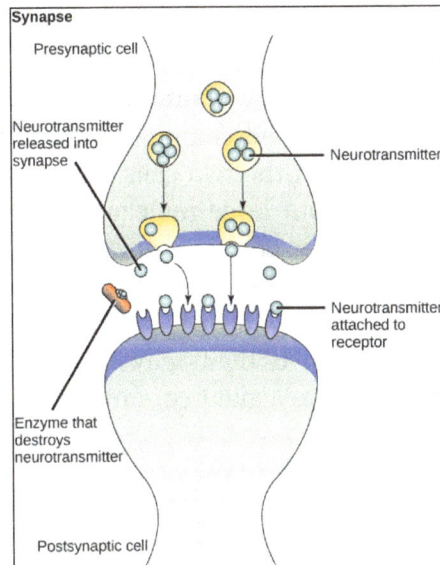

Synapsis: The distance between the presynaptic cell and the postsynaptic cell—called the synaptic gap—is very small and allows for rapid diffusion of the neurotransmitter. Enzymes in the synapatic cleft degrade some types of neurotransmitters to terminate the signal.

Endocrine Signaling

Signals from distant cells are called endocrine signals; they originate from endocrine cells. In the body, many endocrine cells are located in endocrine glands, such as the thyroid gland, the hypothalamus, and the pituitary gland. These types of signals usually produce a slower response, but have a longer-lasting effect. The ligands released in endocrine signaling are called hormones, signaling molecules that are produced in one part of the body, but affect other body regions some distance away.

Hormones travel the large distances between endocrine cells and their target cells via the bloodstream, which is a relatively slow way to move throughout the body. Because of their form of transport, hormones get diluted and are present in low concentrations when they act on their target cells. This is different from paracrine signaling in which local concentrations of ligands can be very high.

Autocrine Signaling

Autocrine signals are produced by signaling cells that can also bind to the ligand that is released. This means the signaling cell and the target cell can be the same or a similar cell (the prefix auto- means self, a reminder that the signaling cell sends a signal to itself). This type of signaling often occurs during the early development of an organism to ensure that cells develop into the correct tissues and take on the proper function. Autocrine signaling also regulates pain sensation and inflammatory responses. Further, if a cell is infected with a virus, the cell can signal itself to undergo programmed cell death, killing the virus in the process. In some cases, neighboring cells of the same type are also influenced by the released ligand. In embryological development, this process of stimulating a group of neighboring cells may help to direct the differentiation of identical cells into the same cell type, thus ensuring the proper developmental outcome.

Direct Signaling across Gap Junctions

Gap junctions in animals and plasmodesmata in plants are connections between the plasma membranes of neighboring cells. These water-filled channels allow small signaling molecules, called intracellular mediators, to diffuse between the two cells. Small molecules, such as calcium ions (Ca^{2+}), are able to move between cells, but large molecules, like proteins and DNA, cannot fit through the channels. The specificity of the channels ensures that the cells remain independent, but can quickly and easily transmit signals. The transfer of signaling molecules communicates the current state of the cell that is directly next to the target cell; this allows a group of cells to coordinate their response to a signal that only one of them may have received. In plants, plasmodesmata are ubiquitous, making the entire plant into a giant communication network.

Types of Receptors

Receptors, either intracellular or cell-surface, bind to specific ligands, which activate numerous cellular processes. Receptors are protein molecules in the target cell or on its surface that bind ligands. There are two types of receptors: internal receptors and cell-surface receptors.

Internal Receptors

Internal receptors, also known as intracellular or cytoplasmic receptors, are found in the cytoplasm of the cell and respond to hydrophobic ligand molecules that are able to travel across the plasma membrane. Once inside the cell, many of these molecules bind to proteins that act as regulators of mRNA synthesis to mediate gene expression. Gene expression is the cellular process of transforming the information in a cell's DNA into a sequence of amino acids that ultimately forms a protein. When the ligand binds to the internal receptor, a conformational change exposes a DNA-binding site on the protein. The ligand-receptor complex moves into the nucleus, binds to specific regulatory regions of the chromosomal DNA, and promotes the initiation of transcription. Internal receptors can directly influence gene expression without having to pass the signal on to other receptors or messengers.

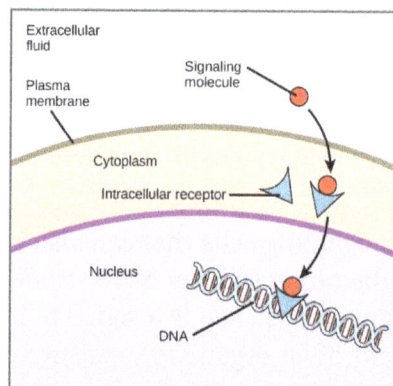

Intracellular Receptors: Hydrophobic signaling molecules typically diffuse across the plasma membrane and interact with intracellular receptors in the cytoplasm. Many intracellular receptors are transcription factors that interact with DNA in the nucleus and regulate gene expression.

Cell-surface Receptors

Cell-surface receptors, also known as transmembrane receptors, are cell surface, membrane-anchored, or integral proteins that bind to external ligand molecules. This type of receptor spans the plasma membrane and performs signal transduction, converting an extracellular signal into an intracellular signal. Ligands that interact with cell-surface receptors do not have to enter the cell that they affect. Cell-surface receptors are also called cell-specific proteins or markers because they are specific to individual cell types.

Each cell-surface receptor has three main components: an external ligand-binding domain (extracellular domain), a hydrophobic membrane-spanning region, and an intracellular domain inside the cell. The size and extent of each of these domains vary widely, depending on the type of receptor.

Cell-surface receptors are involved in most of the signaling in multicellular organisms. There are three general categories of cell-surface receptors: ion channel-linked receptors, G-protein-linked receptors, and enzyme-linked receptors.

Ion Channel-linked Receptors

Ion channel-linked receptors bind a ligand and open a channel through the membrane that allows specific ions to pass through. To form a channel, this type of cell-surface receptor has an extensive membrane-spanning region. In order to interact with the phospholipid fatty acid tails that form the center of the plasma membrane, many of the amino acids in the membrane-spanning region are hydrophobic in nature. Conversely, the amino acids that line the inside of the channel are hydrophilic to allow for the passage of water or ions. When a ligand binds to the extracellular region of the channel, there is a conformational change in the protein's structure that allows ions such as sodium, calcium, magnesium, and hydrogen to pass through.

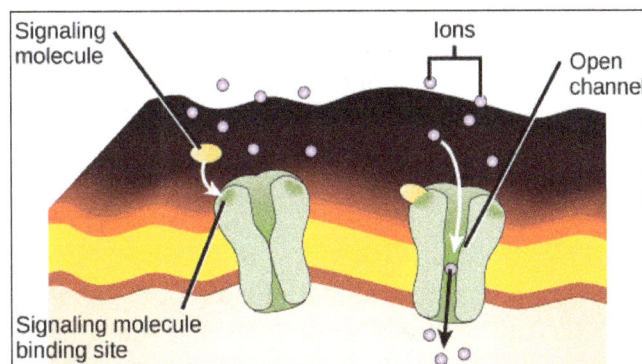

Gated-Ion Channels: Gated ion channels form a pore through the plasma membrane that opens when the signaling molecule binds. The open pore then allows ions to flow into or out of the cell.

G-protein Linked Receptors

G-protein-linked receptors bind a ligand and activate a membrane protein called a G-protein. The activated G-protein then interacts with either an ion channel or an enzyme in the membrane. All G-protein-linked receptors have seven transmembrane domains, but each receptor has its own specific extracellular domain and G-protein-binding site.

Cell signaling using G-protein-linked receptors occurs as a cyclic series of events. Before the ligand binds, the inactive G-protein can bind to a newly-revealed site on the receptor specific for its binding. Once the G-protein binds to the receptor, the resultant shape change activates the G-protein, which releases GDP and picks up GTP. The subunits of the G-protein then split into the α subunit and the β subunit. One or both of these G-protein fragments may be able to activate other proteins as a result. Later, the GTP on the active α subunit of the G-protein is hydrolyzed to GDP and the β subunit is deactivated. The subunits reassociate to form the inactive G-protein, and the cycle starts over.

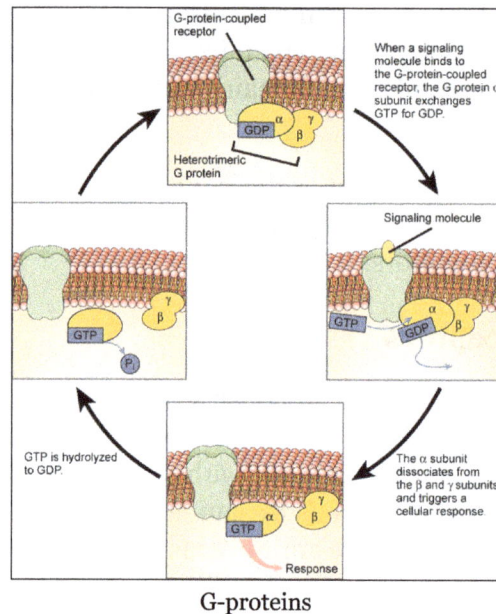

G-proteins

In figure, Heterotrimeric G proteins have three subunits: α, β, and γ. When a signaling molecule binds to a G-protein-coupled receptor in the plasma membrane, a GDP molecule associated with the α subunit is exchanged for GTP. The β and γ subunits dissociate from the α subunit, and a cellular response is triggered either by the α subunit or the dissociated β pair. Hydrolysis of GTP to GDP terminates the signal.

Enzyme-linked Receptors

Enzyme-linked receptors are cell-surface receptors with intracellular domains that are associated with an enzyme. In some cases, the intracellular domain of the receptor itself is an enzyme or the enzyme-linked receptor has an intracellular domain that interacts directly with an enzyme. The enzyme-linked receptors normally have large extracellular and intracellular domains, but the membrane-spanning region consists of a single alpha-helical region of the peptide strand. When a ligand binds to the extracellular domain, a signal is transferred through the membrane and activates the enzyme, which sets off a chain of events within the cell that eventually leads to a response. An example of this type of enzyme-linked receptor is the tyrosine kinase receptor. The tyrosine kinase receptor transfers phosphate groups to tyrosine molecules. Signaling molecules bind to the extracellular domain of two nearby tyrosine kinase receptors, which then dimerize. Phosphates are then added to tyrosine residues on the intracellular domain of the receptors and can then transmit the signal to the next messenger within the cytoplasm.

Signaling Molecules

Signaling molecules are necessary for the coordination of cellular responses by serving as ligands and binding to cell receptors. Produced by signaling cells and the subsequent binding to receptors in target cells, ligands act as chemical signals that travel to the target cells to coordinate responses. The types of molecules that serve as ligands are incredibly varied and range from small proteins to small ions like calcium (Ca^{2+}).

Small Hydrophobic Ligands

Small hydrophobic ligands can directly diffuse through the plasma membrane and interact with internal receptors. Important members of this class of ligands are the steroid hormones. Steroids are lipids that have a hydrocarbon skeleton with four fused rings; different steroids have different functional groups attached to the carbon skeleton. Steroid hormones include the female sex hormone, estradiol, which is a type of estrogen; the male sex hormone, testosterone; and cholesterol, which is an important structural component of biological membranes and a precursor of steriod hormones. Other hydrophobic hormones include thyroid hormones and vitamin D. In order to be soluble in blood, hydrophobic ligands must bind to carrier proteins while they are being transported through the bloodstream.

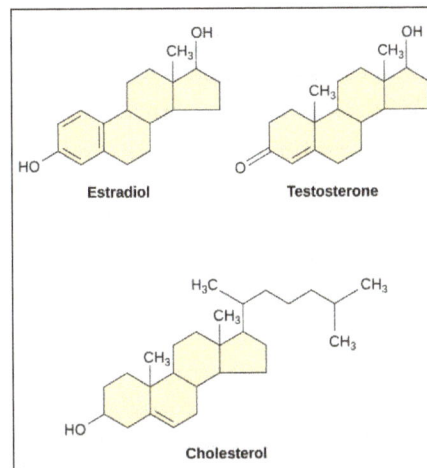

Steroid Hormones: Steroid hormones have similar chemical structures to their precursor, cholesterol. Because these molecules are small and hydrophobic, they can diffuse directly across the plasma membrane into the cell, where they interact with internal receptors.

Water-soluble Ligands

Water-soluble ligands are polar and, therefore, cannot pass through the plasma membrane unaided; sometimes, they are too large to pass through the membrane at all. Instead, most water-soluble ligands bind to the extracellular domain of cell-surface receptors. Cell-surface receptors include: ion-channel, G-protein, and enzyme-linked protein receptors. The binding of these ligands to these receptors results in a series of cellular changes. These water soluble ligands are quite diverse and include small molecules, peptides, and proteins.

Other Ligands

Nitric oxide (NO) is a gas that also acts as a ligand. It is able to diffuse directly across the plasma membrane; one of its roles is to interact with receptors in smooth muscle and induce relaxation of the tissue. NO has a very short half-life; therefore, it only functions over short distances. Nitroglycerin, a treatment for heart disease, acts by triggering the release of NO, which causes blood vessels to dilate (expand), thus restoring blood flow to the heart.

Cell Signaling Pathways

Cell signaling pathways can be generally categorized into groups based on area of biology.

PI3K-AKT Signaling Pathway

PI3K-Akt Pathway is an intracellular signal transduction pathway that promotes metabolism, proliferation, cell survival, growth and angiogenesis in response to extracellular signals. This is mediated through serine and threonine phosphorylation of a range of downstream substrates. Key proteins involved are phosphatidylinositol 3-kinase (PI3K) and Akt/Protein Kinase B.

The origins of PKB/Akt research can be traced back to the discovery in 1977, by Staal and co-workers, which is a previously undescribed oncogene in a virus termed ATK8. And this cell-derived oncogenic sequence were isolated and named *akt*. In 1991, three independent research teams identified genes corresponding to PKB/Akt. These three cloning papers established PKB/Akt as a novel phospho-protein kinase that was widely expressed, and paved the way for future experiments into the role of PKB/Akt in diverse cellular processes. An enzyme termed phosphatidylinositol 3-kinase (PI3K) had been isolated in 1990 by the group of Cantley. In 1995 Richard Roth and his colleagues reported that Akt was activated by insulin. Then, several researches suggested that membrane phospholipids generated by PI3K were an integral element required for PKB/Akt activation.

Composition of PI3K-AKT Pathway

The key molecules involved in this signaling pathway are receptor tyrosine kinase (RTKs), phosphatidylinositol 3-kinase (PI3K), phosphatidylinositol-4,5-bisphosphate (PIP2), phosphatidylinositol-3,4,5-bisphosphate (PIP3) and AKT/protein kinase B.

RTKs are the high-affinity cell surface receptors for many polypeptide growth factors, cytokines, and hormones. This receptor has three functional domains: an extracellular ligand binding domain, a transmembrane domain and an intracellular tyrosine kinase domain. When the ligands such as growth factor bind to the RTKs, two RTKs monomer get close and form a dimer, which leads to activation of the intracellular tyrosine kinase domain and auto phosphorylation by each monomers.

PI3K is a kinase that capable of phosphorylating the 3 position hydroxyl group of the inositol ring of phosphatidylinositol. PI3K consisted of two domains: a catalytic domain P110 and a regulatory domain P85. The activation of PI3K typically occurs as a result of directly stimulated via the regulatory subunit bound to the activated receptor or indirectly activated via adapter molecules such as the insulin receptor substrate (IRS) proteins. PI3K can also be activated by a GTP binding RAS protein.

PIP2 and PIP3 are minor phospholipid components of cell membranes. The PIP3 usually serves as docking phospholipids that bind specific domains that promote the recruitment of proteins to the plasma membrane and subsequent activation of signaling cascades. In PI3K-AKT pathway, the 3 position phosphate group of PIP3 can bind to both PDK1 and AKT protein and recruiting AKT protein at the plasma membrane, allowing PDK1 to access and phosphorylate T308 in the "activation loop", leading to partial PKB/Akt activation. Then the phosphorylation of Akt at S473 in the carboxy-terminal hydrophobic motif, either by mTORC2 or by DNA-PK, stimulates full Akt activity.

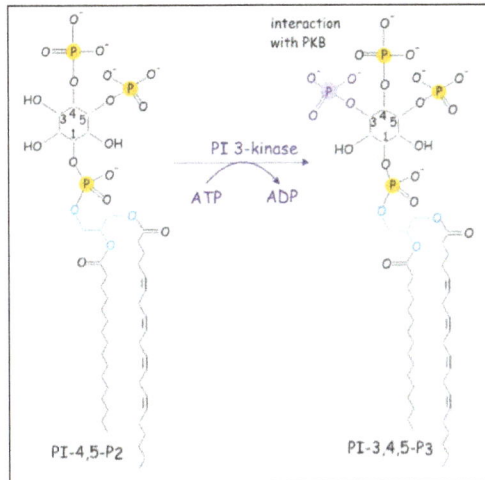

Phosphoinositide metabolism and the structure of PIP2 and PIP3

AKT also named as protein kinase B, is a serine/threonine-specific protein kinase that plays a key role in multiple cellular processes. Once activated, Akt regulates the function via phosphorylation activation or suppression of a broad array of proteins involved in cell growth, proliferation, motility, adhesion, neovascularization, and cell death.

Regulation of PI3K-AKT Signaling

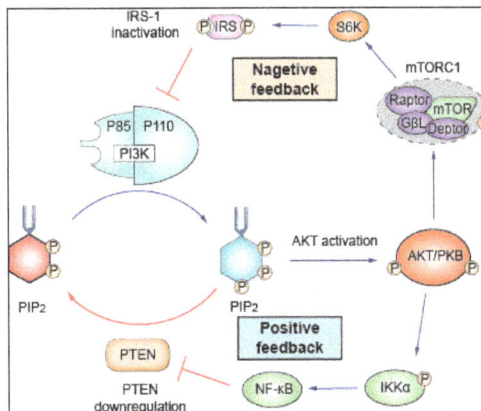

The positive and negative feedback mechanism of PI3K-AKT pathway

The PI3K-Akt pathway has many downstream effects and must be carefully regulated. Negatively regulation of PI3K-AKT pathway can be achieve at to target: the PIP3 level and the inactivation of AKT protein. Phosphatase and tensin homolog (PTEN) is a main down regulation protein which can converting PIP3 into PIP2. Protein phosphatase 2A (PP2A), which dephosphorylates Akt at

Thr308 and phosphatase PHLPP dephosphorylates Akt at Ser473 are also two negative regulation proteins. Addition to these regulation protein, the pathway itself also have feedback mechanisms: Transcription factor NF-κB, activated by Akt, regulates peroxisome proliferator-activated receptor delta (PPARβ/δ) agonists and tumor necrosis factor α (TNFα), which in turn repress PTEN expression as a positive feedback; Negative feedback loop is functioned by mTORC1 and S6K1 activation. S6K1 is able to phosphorylate IRS-1 at multiple serine residues, preventing binding to RTKs, resulting the suppression of PI3K activation.

Downstream of PI3K-AKT Signaling

Once active, Akt could translocated from the plasma membrane to the cytoplasm and nucleus, where many of its substrates reside. Phosphorylation by Akt can be inhibitory or stimulatory, either suppressing or enhancing the activity of target proteins. Depending on the target protein, Akt can regulate different cell functions, here we discuss several main effect of PI3K-Akt pathway:

- Cell apoptosis/Survival: Akt enhances the survival of cells by blocking the function of proapoptotic proteins and processes. Akt negatively regulates the function or expression of Bcl-2 family members, Bax protein, and Bim protein. Akt also inhibits the expression of BH3-only proteins through effects on transcription factors, such as FOXO and p53. p53 is also an oncogene that mediate cell apoptosis. Akt can promote the p53 degradation through phosphorylation of MDM2. Akt can also phosphorylates GSK3 isoforms on a highly conserved N-terminal regulatory site, and inactivates the kinase, so that regulate the apoptosis and glucose metabolism via GSK3.

- Protein synthesis and cell growth: One of the best-conserved functions of Akt is its role in promoting cell growth through inhibition of TSC2 and indirectly activation of mTOR complex 1 (mTORC1). mTORC1 is a critical regulator of translation initiation and ribosome biogenesis and plays an evolutionarily conserved role in cell growth control. It can activate S6K and eukaryotic initiation factor 4E (eIF4E)-binding protein 1 (4E-BP1), S6K can activate ribosomal S6 and promote the protein synthesis and cell growth.

- Cell Proliferation and cell cycle: The function of P21/Waf1/Cip1 protein is to maintain the cell in the quiescent state. And P27/Kip2 have a similar function to maintain the cell in G1 state. Protein Akt can also phosphorylates P21/Waf1/Cip1 and P27/Kip2, and inhibits their anti-proliferative effects by retaining them in the cytoplasm. So it can promote the cell into cell cycle to proliferation.

In general, the function of PI3K-AKT signaling pathway is to stimulate cell to proliferation and growth, and simultaneously inhibit cell apoptosis.

Relationship with Oncogenesis

As the general regulation direction of PI3K-AKT pathway is to stimulate the cell growth and proliferation. The over activation of this signaling pathway can over stimulate cell and resulting abnormal cell proliferation, that is oncogenesis. Abnormalities in the PI3K pathway are common in cancer and have a role in neoplastic transformation. PI3K itself is a frequent target of mutational activation. The most frequent genetic aberrations in breast cancer are somatic missense mutations in the PIK3CA gene encoding p110α. There are two kinds of mutations of PIK3CA gene, abnormalities of gene copy number and point mutation that resulting the persistently activation of PI3K without the RTKs stimulation. 80% of the point mutation of PI3K are E542K, E545K and H1047R mutation. The patients with these point mutation are insensitive to the targeting anticancer drugs of EGFR and Her2 inhibitor such as Lapatinib and Cetuximab.

Table: Abnormalities in the PI3K/AKT signaling pathway in cancer.

Molecule	Alteration in Tumors	Frequency	Tumour lineage
PTEN	Mutations (somatic)	>50%	Glioma, melanoma, prostate cancer Endometrial cancer, endometrioid ovarian cancer Variable in sporadic breast cancers (2–30%).
	Decreased expression Methylation Loss of heterozygosity	>50%	Breast, melanoma, prostate Microsatellite instability-high colorectal cancer Endometrial cancer Leukaemia.
	Germline mutations	80% of Cowden's disease	High risk of breast, thyroid and endometrial carcinomas.
p85	Activating mutations	Rare	Ovary, colon, glioma, lymphoma cell line (CO)
	Fusion	Very rare	Lymphoma
PIK3CA	Amplification	Up to 50% Rare	Ovary, cervix, lung Breast (BRCA1 associated?)
	Activating mutation	>50% >25%	Bowel Breast
AKT1	Amplification	Low	Gastric
AKT2	Amplification	Low	Ovary (12–25%) Pancreas (20%), breast (rare)
	Mutation	Low	Colorectal
AKT3	Overexpression	Low	Hormone-resistant prostate and breast cancer.
PDK1	Mutation	Low	Colorectal

AKT and PTEN are also targets of frequent genomic and epigenetic change in human cancers. The most common mutation of gene which expresses AKT are activation point mutation and abnormal increasing of gene copy number. The activation mutation of E17L and Q79K are very frequently observed in breast cancer, colon cancer, ovarian cancer and carcinoma of urinary bladder. The common mutation of PTEN gene are abnormal loss of gene copy number and inactivation point mutation, over 10% of patients with endometrial cancer, brain cancer, skin cancer and prostate cancer have PTEN gene mutation.

Several small molecules that inhibit the PI3K–Akt signaling pathway are in clinical development. Four main classes of inhibitors are usually discussed: dual PI3K–mTOR inhibitors, PI3K inhibitors, Akt inhibitors and mTOR inhibitors.

Table: PI3K–Akt pathway inhibitors in clinical development for treating cancers.

Inhibitor	Company	Phase of clinical trial
Dual PI3K and mTOR inhibitors		
BEZ235	Novartis	Phase I/II
BGT226	Novartis	Phase I/II
XL765	Exelixis	Phase I

Inhibitor	Company	Phase of clinical trial
SF1126	Semafore	Phase I/II
GSK1059615	GSK	Preclinical
PI3K inhibitors		
XL147	Exelixis	Phase I
PX866	Oncothyreon	Phase I
GDC0941	Genentech/Piramed/Roche	Phase I
BKM120	Novartis	Phase I
CAL101 (targets p110δ)	Calistoga Pharmaceuticals	Phase I
Akt inhibitors		
Perifosine	Keryx	Phase I/II
GSK690693	GSK	Phase I
VQD002	Vioquest	Phase I
MK2206	Merck	Phase I
mTOR inhibitors (catalytic site)		
OSI027	OSI Pharmaceuticals	Phase I
AZD8055	AstraZeneca	Phase I/II

JNK Signaling Pathway

The c-Jun N-terminal kinase (JNK) pathway is one of the major signaling cassettes of the mitogen-activated protein kinase (MAPK) signaling pathway. It functions in the control of a number of cellular processes, including proliferation, embryonic development and apoptosis. The pathway takes its name from the c-Jun N-terminal kinases 1–3 (JNK1–JNK3), which are the MAPKs that interact with the final effectors. The JNK pathway is activated by a bewildering number of mechanisms. This complexity is evident by the fact that there are 13 MAPK kinase kinases (MAPKKKs) responsible for feeding information into the JNK pathway. The JNK pathway can also be activated through G protein-coupled receptors (GPCRs) using G proteins such as G12/13.

JNK pathway contributes to the control of a large number of cellular processes:

- Phosphorylation of the transcription factor p53.

- The JNK pathway has been implicated in the mitogen activated protein kinase (MAPK) signaling in cardiac hypertrophy.

The c-Jun N-terminal kinases (JNKs) were first described in the early 1990s. Landmark discoveries in the area of JNK research are summarized in figure.

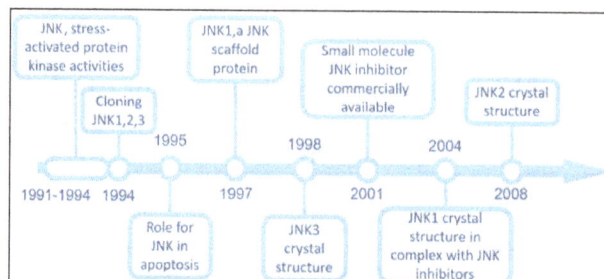

Landmark discoveries in the area of JNK research

JNK Signaling Cascade

The JNK module plays an important role in apoptosis, inflammation, cytokine production, and metabolism. The JNK module is activated by environmental stresses (ionizing radiation, heat, oxidative stress, and DNA damage) and inflammatory cytokines, as well as growth factors, and signaling to the JNK module often involves the Rho family GTPases Cdc42 and Rac. Those receptors or receptor-independent stress-induced membranal changes further transmit the signals to adaptor proteins that can by themselves activate kinases in the MAP4K, and sometime, MAP3K tiers of the JNK cascade. Next, the activated MAP3Ks transmit the signals to kinases at the MAPKK level, which are mainly MKK4 and MKK7. As the other MAPKKs, the main JNK kinases (MKK4, MKK7) are activated by phosphorylation of the typical Ser-Xaa-Ala-Xaa-Ser/Thr motif (Ser 198, Thr 202 in MKK7) and are then able to transmit the signal further to the JNK level. Shortly after activation, like the other MAPKs, JNKs translocate into the nucleus where they usually physically associate with their targets- transcription factors such as c-Jun, ATF, Elk1 and activate them.

- MEK4: The mitogen-activated protein kinase kinase 4 (MKK4) was first identified in a cDNA library from Xenopus laevis embryos by a PCR based screen and called XMEK2. The mkk4 gene is located on human chromosome 17, encodes a protein of 399 amino acids and shares 94% homology with the mouse and rat protein. MKK4 is involved in a variety of physiological and pathophysiological processes.

- MEK7: The mitogen-activated protein kinase kinase 7 (MKK7), also called stress-activated protein kinase/extracellular signal regulated protein kinase kinase 2 (SEK2) or c-Jun N-terminal kinase kinase 2 (JNKK2) was first cloned from cDNA of murine testis tissue. The loss of mkk7 leads to decreased proliferation.

- JNKs: The c-Jun N-terminal kinases (JNKs), also called stress-activated protein kinases (SAPKs), have first been discovered in a c-Jun binding assay with extracts from UV-stimulated HeLa cells. JNKs are activated by concomitant phosphorylation of a threonine and a tyrosine residue within a conserved T-P-Y motif in the activation loop of the kinase domain by MKK4 and MKK7. The jnk1 and jnk2 are widely expressed, whereas expression of jnk3 is mainly confined to brain, heart and testis.

Downstream Signaling of JNK Pathway

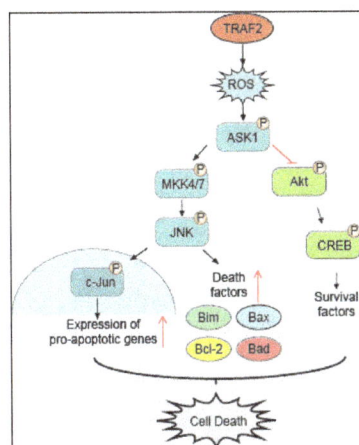

Downstream signaling of JNK pathway

JNK pathway is generally a "death" signaling pathway. It control the cell response to the harmful extracellular stimulus such as inflammatory cytokines, UV-irradiation gamma-irradiation etc.. Cells under these harmful stimulus may have DNA mutation or damage. If the DNA damage could not be repaired immediately, the cell must be programed to death (also called apoptosis) to avoid these bad mutation or damage. JNK signaling pathway is such a death pathway that control cell death. There are two main downstream signaling of JNK pathway: one is activation of death signaling such c-Jun, Fos and apoptosis signaling such as BIM, BAD, BAX protein or active P53 transcription, to promote cell apoptosis; the other is inhibition of the cell survival signaling such as STATs and CREB.

JNK can promote apoptosis by two distinct mechanisms. In the first mechanism targeted at the nuclear events, activated JNK translocates to the nucleus and transactivates c-Jun and other target transcription factors (TF). JNK can promote apoptosis by increasing the expression of pro-apoptotic genes through the transactivation of c-Jun/AP1-dependent or p53/73 protein-dependent mechanisms. In pathways directed at mitochondrial apoptotic proteins, activated JNK translocates to mitochondria. There, JNK can phosphorylate the BH3-only family of Bcl2 proteins to antagonize the anti-apoptotic activity of Bcl2 or Bcl-XL. In addition, JNK can stimulate the release of cytochrome c (Cyt C) from the mitochondrial inner membrane through a Bid-Bax-dependent mechanism, promoting the formation of apoptosomes consisting of cytochrome c, caspase-9 (Casp 9) and Apaf-1. This complex initiates the activation of caspase-9-dependent caspase cascade. In another mechanism, JNK can promote the release of Smac/Diablo (Smac) that can inhibit the TRAF2/IAP1 inhibitory complex, thereby relieving the inhibition on caspase-8 to initiate caspase activation. In addition, by phosphorylating BAD and its sequestering partner 14-3-3, JNK can promote BAD-mediated neutralization of the Bcl2 family of anti-apoptotic proteins. Finally, JNK can phosphorylate Bcl2 for suppressing its anti-apoptotic activity.

Regulation of JNK signaling

- Scaffold proteins in the JNK pathway: Just like the other MAP kinase signaling pathway, JNK pathway could also be regulated by the scaffold proteins. Five structurally distinct scaffold proteins have been shown to assemble the JNK signaling unit. These proteins contain a variety of protein-interacting motifs and constitute a family normally referred to as JIPs. Scaffold proteins can tethering individual pathway components together, so that different pathways can be insulated from one another. This becomes particularly useful when single components are used alternatively in different pathways. Scaffold proteins can also establish connections between pathways and help to distribute signals. More recent modelling data suggest that depending on cellular conditions, spatial organization of kinases on scaffold proteins can either enhance or inhibit signal propagation through a kinase cascade.

- Phosphatases in JNK pathway: To date, the reported JNK phosphatases are the dual-specificity MKPs. MKP-7 and VH5 are JNK phosphatases that also dephosphorylate p38a and b. Other MKPs with some reported activity toward JNKs are DSP2, MKP6, MKP1, MKP2, MKP4 and MKP5. Though little is known about the phosphatases that act on upstream MAP2Ks and MAP3Ks, the phosphoserine/threonine protein phosphatase 5 (PP5) was reported to suppress hypoxia-induced ASK1/MEK4/JNK signaling.

- Other modifications affecting JNK activation: JNK activation is also regulated by ubiquitylation. Activation of MEKK1 not only stimulates its ability to phosphorylate and activate MEK1 and MEK4 in the ERK1/2 and JNK pathways but also leads to the ubiquitylation of MEKK1 itself. This ubiquitylation subsequently inhibits MEKK1-catalysed phosphorylation of MEK1 and MEK4, thereby downregulating the MEKK1 activation of the ERK1/2 and JNK pathways.

Relationship with Diseases

- JNK in cancer: JNK is implicated in oncogenic transformation; A role for the JNK pathway in tumorigenesis is supported by the high levels of JNK activity found in several cancer cell lines. The most insightful evidence for a role of JNK/SAPK signaling modules in cancer has come from the identification of MKK4 as a putative tumor suppressor. MKK4 is a MKK capable of phosphorylating both JNKs and p38 MAPKs. This distinguishes MKK4 from MKK7 which has only been shown to phosphorylate JNKs and not p38 MAPKs. Genetic inactivation of the MKK4 gene on chromosome 17 p has been reported for several different tumor types including a subset of breast, biliary and pancreatic carcinomas.

- JNK in nervous system disease: JNK3 activation is a pro-apoptotic pathway in hippocampal neurons. Consistent with a role for JNK3 in excitotoxic neuronal cell death. The role of JNK3 in excitotoxicity suggested that JNK/SAPK pathways would be involved in other neuronal death responses and neurodegenerative diseases. A clear role of JNK/ SAPKs in ischemia-induced cell death has been demonstrated in mice. JNK3 may also prove to be a therapeutic target for neurodegenerative diseases including Parkinson's and Alzheimer's disease.

JNK/SAPKs are clearly involved in ischemia-induced cell death and reperfusion injury in several different tissues and the control of insulin sensitivity in metabolic regulation. There are many other suggestions in the literature that link JNK/SAPK signaling to additional human diseases such as type I diabetes, osteosarcoma, ataxia and immune system dysfunction. JNKs probably play a role in chronic inflammation, airway hyper responsiveness and protease-directed tissue remodeling. It is likely that selective JNK1, JNK2 and JNK3 inhibitors will be needed for specificity and lack of toxicity. It may also be useful to develop specific MKKK inhibitors to selectively block JNK activation in response to different upstream inputs.

Signal Relay Pathways

The chains of molecules that relay signals inside a cell are known as intracellular signal transduction pathways. Here, we'll look at the general characteristics of intracellular signal transduction pathways, as well as some relay mechanisms commonly used in these pathways.

Binding Initiates a Signaling Pathway

When a ligand binds to a cell-surface receptor, the receptor's intracellular domain (part inside the cell) changes in some way. Generally, it takes on a new shape, which may make it active as an enzyme or let it bind other molecules.

The change in the receptor sets off a series of signaling events. For instance, the receptor may turn on another signaling molecule inside of the cell, which in turn activates its own target. This chain reaction can eventually lead to a change in the cell's behavior or characteristics, as shown in the figure below.

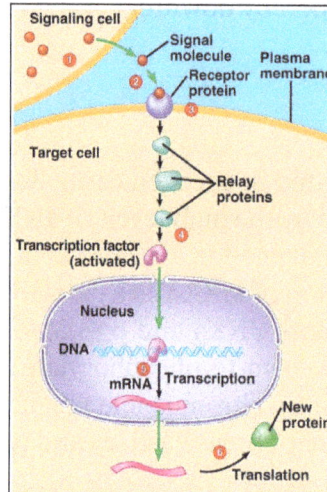

How the components of a hypothetical signaling pathway are activated sequentially, with one turning on the next to produce a cellular response.

Because of the directional flow of information, the term upstream is often used to describe molecules and events that come earlier in the relay chain, while downstream may be used to describe those that come later (relative to a particular molecule of interest). For instance, in the diagram, the receptor is downstream of the ligand but upstream of the the proteins in the cytosol. Many signal transduction pathways amplify the initial signal, so that one molecule of ligand can lead to the activation of many molecules of a downstream target.

The molecules that relay a signal are often proteins. However, non-protein molecules like ions and phospholipids can also play important roles.

Phosphorylation

The figure features a bunch of blobs (signaling molecules) labeled as "on" or "off." What does it actually mean for a blob to be on or off? Proteins can be activated or inactivated in a variety of ways. However, one of the most common tricks for altering protein activity is the addition of a phosphate group to one or more sites on the protein, a process called phosphorylation.

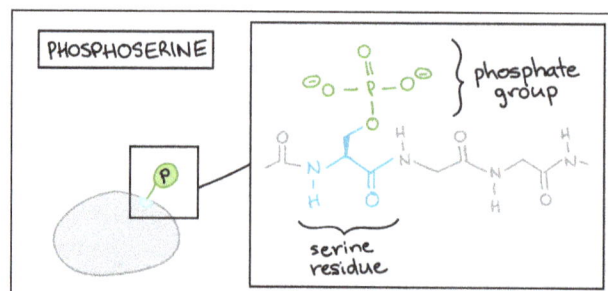

Phosphorylated protein bearing a phosphate group attached to a serine residue, showing the actual chemical structure of the linkage.

Phosphate groups can't be attached to just any part of a protein. Instead, they are typically linked to one of the three amino acids that have hydroxyl (-OH) groups in their side chains: tyrosine, threonine, and serine. The transfer of the phosphate group is catalyzed by an enzyme called a kinase, and cells contain many different kinases that phosphorylate different targets.

Phosphorylation often acts as a switch, but its effects vary among proteins. Sometimes, phosphorylation will make a protein more active (for instance, increasing catalysis or letting it bind to a partner). In other cases, phosphorylation may inactivate the protein or cause it to be broken down.

In general, phosphorylation isn't permanent. To flip proteins back into their non-phosphorylated state, cells have enzymes called phosphatases, which remove a phosphate group from their targets.

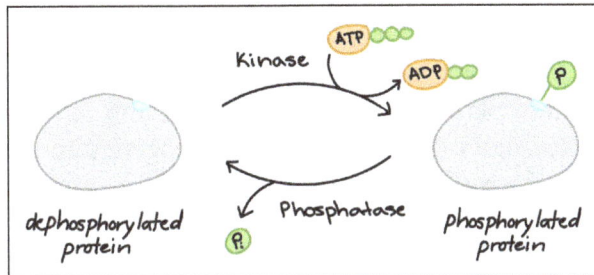

In figure, showing how a protein is phosphorylated by a kinase through the addition of a phosphate from ATP, producing ADP as a by-product, and dephosphorylated by a phosphatase, releasing Pi (inorganic phosphate) as a by-product. The two reactions make up a cycle in which the protein toggles between two states.

Phosphorylation Example: MAPK Signaling Cascade

To get a better sense of how phosphorylation works, let's examine a real-life example of a signaling pathway that uses this technique: growth factor signaling. Specifically, we'll look at part of the epidermal growth factor (EGF) pathway that acts through a series of kinases to produce a cellular response.

This diagram shows part of the epidermal growth factor signaling pathway.

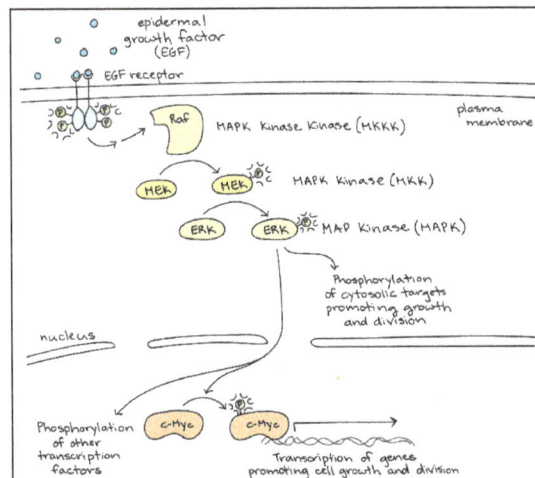

Phosphorylation (marked as a P) is important at many stages of this pathway:

- When growth factor ligands bind to their receptors, the receptors pair up and act as kinases, attaching phosphate groups to one another's intracellular tails.

- The activated receptors trigger a series of events. These events activate the kinase Raf.

- Active Raf phosphorylates and activates MEK, which phosphorylates and activates the ERKs.

- The ERKs phosphorylate and activate a variety of target molecules. These include transcription factors, like c-Myc, as well as cytoplasmic targets. The activated targets promote cell growth and division.

Together, Raf, MEK, and the ERKs make up a three-tiered kinase signaling pathway called a mitogen-activated protein kinase (MAPK) cascade. (A *mitogen* is a signal that causes cells to undergo *mitosis*, or divide.) Because they play a central role in promoting cell division, the genes encoding the growth factor receptor, Raf, and c-Myc are all proto-oncogenes, meaning that overactive forms of these proteins are associated with cancer.

MAP kinase signaling pathways are widespread in biology: they are found in a wide range of organisms, from humans to yeast to plants. The similarity of MAPK cascades in diverse organisms suggests that this pathway emerged early in the evolutionary history of life and was already present in a common ancestor of modern-day animals, plants, and fungi.

Second Messengers

Although proteins are important in signal transduction pathways, other types of molecules can participate as well. Many pathways involve second messengers, small, non-protein molecules that pass along a signal initiated by the binding of a ligand (the "first messenger") to its receptor.

Second messengers include Ca^{2+} ions; cyclic AMP (cAMP), a derivative of ATP; and inositol phosphates, which are made from phospholipids.

Calciumions

Calcium ions are a widely used type of second messenger. In most cells, the concentration of calcium ions (Ca^{2+}) in the cytosol is very low, as ion pumps in the plasma membrane continually work to remove it. For signaling purposes, Ca^{2+} may be stored in compartments such as the endoplasmic reticulum.

In pathways that use calcium ions as a second messenger, upstream signaling events release a ligand that binds to and opens ligand-gated calcium ion channels. These channels open and allow the higher levels of Ca^{2+} that are present outside the cell (or in intracellular storage compartments) to flow into the cytoplasm, raising the concentration of cytoplasmic Ca^{2+}.

How does the released Ca^{2+} help pass along the signal? Some proteins in the cell have binding sites for Ca^{2+} ions, and the released ions attach to these proteins and change their shape (and thus, their activity). The proteins present and the response produced are different in different types of cells. For instance, Ca^{2+} signaling in the β-cells of the pancreas leads to the release of insulin, while Ca^{2+}, signaling in muscle cells leads to muscle contraction.

Cyclic AMP (cAMP)

Another second messenger used in many different cell types is cyclic adenosine monophosphate (cyclic AMP or cAMP), a small molecule made from ATP. In response to signals, an enzyme called adenylyl cyclase converts ATP into cAMP, removing two phosphates and linking the remaining phosphate to the sugar in a ring shape.

Once generated, cAMP can activate an enzyme called protein kinase A (PKA), enabling it to phosphorylate its targets and pass along the signal. Protein kinase A is found in a variety of types of cells, and it has different target proteins in each. This allows the same cAMP second messenger to produce different responses in different contexts.

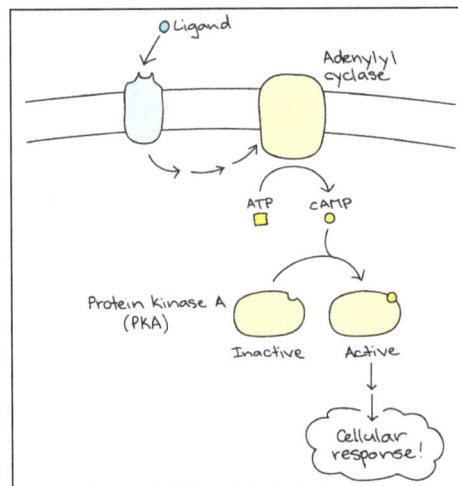

Pathway that uses cAMP as a second messenger

In figure, a ligand binds to a receptor, leading indirectly to activation of adenylyl cyclase, which converts ATP to cAMP. cAMP binds to protein kinase A and activates it, allowing PKA to phosphorylate downstream factors to produce a cellular response.

cAMP signaling is turned off by enzymes called phosphodiesterases, which break the ring of cAMP and turn it into adenosine monophosphate (AMP).

Inositol Phosphates

Although we usually think of plasma membrane phospholipids as structural components of the cell, they can also be important participants in signaling. Phospholipids called phosphatidylinositols can be phosphorylated and snipped in half, releasing two fragments that both act as second messengers.

One lipid in this group that's particularly important in signaling is called PIP_2. In response to a signal, an enzyme called phospholipase C cleaves (chops) PIP_2 two fragments, DAG and IP_3. These fragments made can both act as second messengers.

DAG stays in the plasma membrane and can activate a target called protein kinase C (PKC), allowing it to phosphorylate its own targets. IP_3 diffuses into the cytoplasm and can bind to ligand-gated calcium channels in the endoplasmic reticulum, releasing Ca^{2+} that continues the signal cascade.

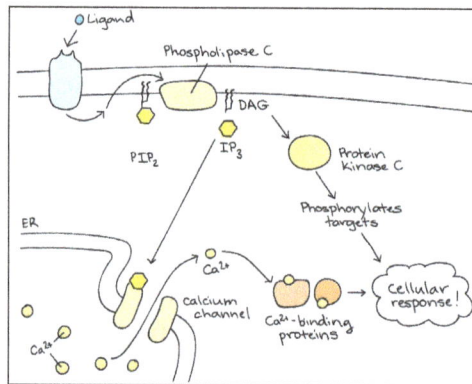

Cell Adhesion and Cell Communication

All cells rely on cell signaling to detect and respond to cues in their environment. This process not only promotes the proper functioning of individual cells, but it also allows communication and coordination among groups of cells — including the cells that make up organized communities called tissues. Because of cell signaling, tissues have the ability to carry out tasks no single cell could accomplish on its own.

Different types of tissues, such as bone, brain, and the lining of the gut, have characteristic features related to the number and types of cells they contain. Cell spacing is also critical to tissue function, so this geometry is precisely regulated. To preserve proper tissue architecture, adhesive molecules help maintain contact between nearby cells and structures, and tiny tunnel-like junctions allow the passage of ions and small molecules between adjacent cells. Meanwhile, signaling molecules relay positional information among the cells in a tissue, as well as between these cells and the extracellular matrix. These signaling pathways are critical to maintaining the state of equilibrium known as homeostasis within a tissue. For example, the processes involved in wound healing depend on positional information in order for normal tissue architecture to be restored. Such positional signals are also crucial for the development of adult structures in multicellular organisms. As tissues develop, clumps of unorganized cells grow and sort themselves according to signals they send and receive.

Promotion of Tissue Structure and Function by Integrins

Within tissues, adhesive molecules allow cells to maintain contact with one another and with structures in the extracellular matrix. One especially important class of adhesive molecules is the integrins. Integrins are more than just mechanical links, however: They also relay signals both to and from cells. In this way, integrins play an important role in sensing the environment and controlling cell shape and motility.

Integrins are a diverse family of transmembrane proteins found in all animal cells. Even simple animals like sponges have these proteins. Each individual integrin consists of two main parts: an alpha subunit and a beta subunit. Variation in the alpha and beta subunits accounts for the wide variety of integrins observed throughout the animal kingdom. For example, humans alone have over 20 different kinds of integrins.

Integrins link the actin cytoskeleton of a cell to various external structures. The cytoplasmic portion of each integrin molecule binds to adaptor proteins that connect to the actin filaments inside the cell. The extracellular portion of the integrin then binds to molecules in the extracellular matrix or on the surface of other cells. Integrin attachments to neighboring cells can break and reform as a cell moves.

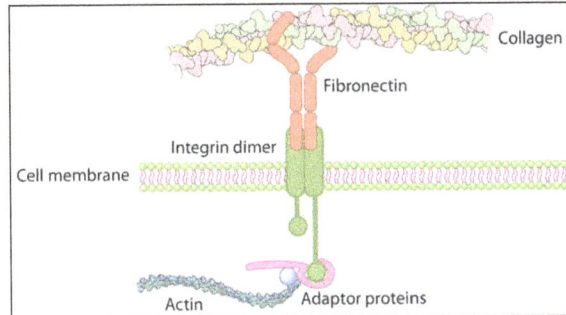

Integrin connects the extracellular matrix with the actin cytoskeleton inside the cell

Contact between Cells within a Tissue

Beyond integrins, cells rely on several other adhesive proteins to maintain physical contact. As an example, consider the epithelial cells that line the inner and outer surfaces of the human body — including the skin, intestines, airway, and reproductive tract. These cells provide a dramatic example of the different kinds of cell-to-cell junctions, but the same junctions also exist in a wide range of other tissues.

The side surfaces of epithelial cells are tightly linked to those of neighboring cells, forming a sheet that acts as a barrier. Within this sheet, each individual cell has a set orientation. Through integrins, the basal end of each cell connects to a specialized layer of extracellular matrix called the basal lamina. In contrast, the apical end of each cell faces out into the environment - such as the inner cavity or lumen of the gut.

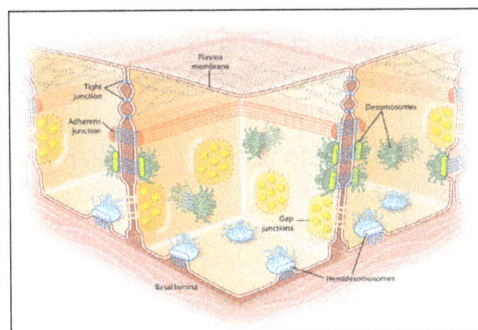

The side-to-side junctions that link epithelial cells are diverse in their protein makeup and function. The adhesive transmembrane proteins anchoring these junctions have extracellular portions that interact with similar proteins on adjacent cells. Protein complexes within each cell further connect the transmembrane adhesive proteins to the cytoskeleton. In particular, adaptor complexes bind adherens junctions to cytoskeletal actin, and other adaptor complexes bind desmosomes to intermediate filaments. Both of these types of junctional complexes provide cells and tissues with mechanical support, and they additionally recruit intracellular signaling molecules to relay positional information to the nucleus.

The different Types of Cell Junctions

In figure, tight junctions (blue dots) between cells are connected areas of the plasma membrane that stitch cells together. Adherens junctions (red dots) join the actin filaments of neighboring cells together. Desmosomes are even stronger connections that join the intermediate filaments of neighboring cells. Hemidesmosomes (light blue) connect intermediate filaments of a cell to the basal lamina, a combination of extracellular molecules on other cell surfaces. Gap junctions (yellow) are clusters of channels that form tunnels of aqueous connectivity between cells.

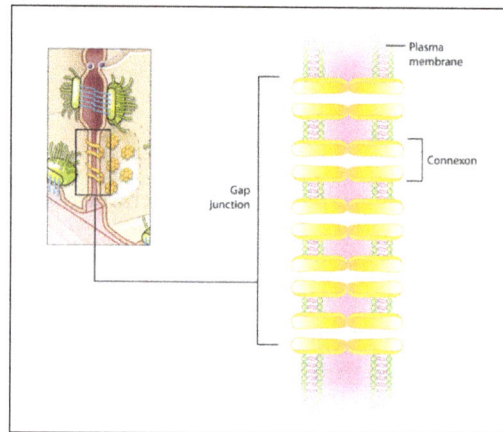

A gap junction: In a gap junction, the lipid bilayer of adjacent cells is pierced through by proteins called connexons. These proteins group together and effectively form a group of communication tunnels between adjacent cells.

The lateral surfaces of epithelial cells also contain several other types of specialized junctions. Tight junctions form a seal between cells that is so strong that not even ions can pass across it. Gap junctions are involved in cellular communication — not just in epithelial tissue, but in other tissue types as well. Gap junctions are specialized connections that form a narrow pore between adjacent cells. These pores permit small molecules and ions to move from one cell to another. In this way, gap junctions provide metabolic and electrical coupling between cells. For example, cardiac tissue has extensive gap junctions, and the rapid movement of ions through these junctions helps the tissue beat in rhythm. Gap junctions may also open and close in response to metabolic signals.

Cell Death can be Prompted by a Signal

Cell signaling isn't just central to tissue architecture and function: It also plays an important role in the balance between cell growth and death. Although it sounds like a bad thing, apoptosis — or the process of programmed cell death — is an essential aspect of development. Without it, repair and replenishment processes would overrun tissues with new cells. The orderly demise of a certain proportion of cells is therefore necessary for normal tissue turnover and maintenance of homeostasis. Apoptosis is distinct from necrosis, a messier form of cell death that causes cells to literally swell and burst. Necrotic cell death is not programmed; rather, it occurs in response to trauma or injury.

A range of extracellular and intracellular signals can trigger either cell growth or apoptosis. When cells receive these signals from their neighbors or from other aspects of the external environment, they carefully weigh them against each other before choosing a course of action. For instance, signals that indicate a lack of nutrients or the presence of toxins would likely stall cell growth and promote apoptosis. Within the cell, damage to the DNA or loss of mitochondrial integrity might also result in programmed cell death.

Cells self-destruct cleanly and quickly during apoptosis, thanks to the activation of a variety of enzymes — proteases and nucleases — that break down proteins and nucleic acids, respectively. In fact, scientists look for a characteristic pattern of fragmentation and nuclear condensation within tissues as evidence that apoptosis has occurred.

References

- Cell-signaling: biologydictionary.net, Retrieved 1 August, 2019

- Three-stages-of-cell-signaling: wordpress.com, Retrieved 28 April, 2019

- Signaling-molecules-and-cellular-receptors, boundless-biology: lumenlearning.com, Retrieved 19 May, 2019

- Pi3k-akt-signaling-pathway: creative-diagnostics.com, Retrieved 28 February, 2019

- Jnk-signaling-pathway: creative-diagnostics.com, Retrieved 18 April, 2019

- Intracellular-signal-transduction, mechanisms-of-cell-signaling, cell-signaling, cell-signaling, science: khanacademy.org, Retrieved 7 May, 2019

- Cell-adhesion-and-cell-communication: nature.com, Retrieved 21 July, 2019

Chapter 5

Biological Activity of Cells

There are several biological activities which take place within cells, such as respiration, reproduction and migration. The metabolic reactions and processes through which cells transform biochemical energy from nutrients into adenosine triphosphate are known as cell respiration. All these diverse biological activities of the cells have been carefully analyzed in this chapter.

Cellular Respiration

Cellular respiration is the process of oxidizing food molecules, like glucose, to carbon dioxide and water.

$$C_6H_{12}O_6 + 6O_2 + 6H_2O \rightarrow 12H_2O + 6\ CO_2$$

The energy released is trapped in the form of ATP for use by all the energy-consuming activities of the cell.

The process occurs in two phases:

- Glycolysis, the breakdown of glucose to pyruvic acid.

- The complete oxidation of pyruvic acid to carbon dioxide and water.

In eukaryotes, glycolysis occurs in the cytosol. The remaining processes take place in mitochondria.

Mitochondria

Mitochondria are membrane-enclosed organelles distributed through the cytosol of most eukaryotic cells. Their number within the cell ranges from a few hundred to, in very active cells, thousands. Their main function is the conversion of the potential energy of food molecules into ATP.

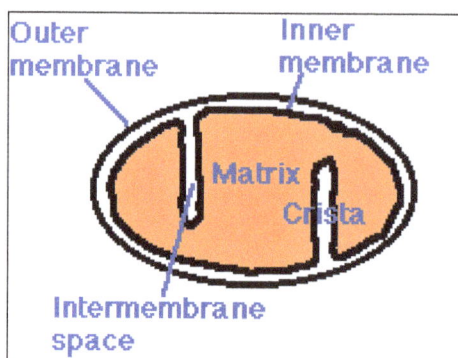

Mitochondria have:

- An outer membrane that encloses the entire structure.

- An inner membrane that encloses a fluid-filled matrix.

- Between the two is the intermembrane space.

- The inner membrane is elaborately folded with shelflike cristae projecting into the matrix.

- A small number (some 5–10) circular molecules of DNA.

This electron micrograph (courtesy of Keith R. Porter) shows a single mitochondrion from a bat pancreas cell. Note the double membrane and the way the inner membrane is folded into cristae. The dark, membrane-bound objects above the mitochondrion are lysosomes.

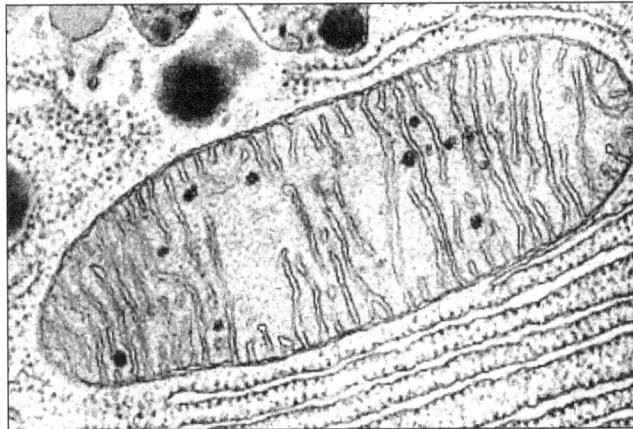

The number of mitochondria in a cell can:

- Increase by their fission (e.g. following mitosis);

- Decrease by their fusing together.

(Defects in either process can produce serious, even fatal, illness.)

Outer Membrane

The outer membrane contains many complexes of integral membrane proteins that form channels through which a variety of molecules and ions move in and out of the mitochondrion.

Inner Membrane

The inner membrane contains 5 complexes of integral membrane proteins:

- NADH dehydrogenase (Complex I),

- Succinate dehydrogenase (Complex II),

- Cytochrome c reductase (Complex III; also known as the cytochrome b-c_1 complex),

- Cytochrome c oxidase (Complex IV),

- ATP synthase (Complex V).

Matrix

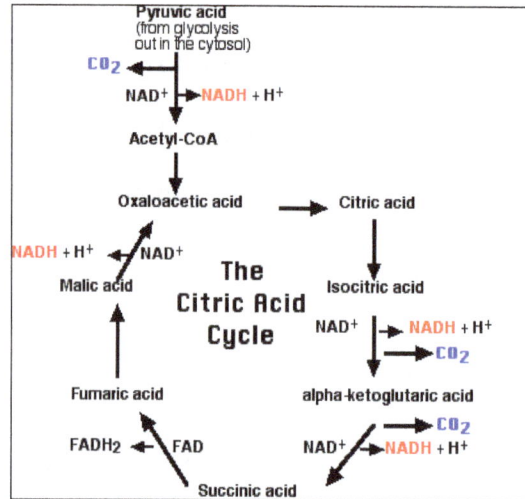

The matrix contains a complex mixture of soluble enzymes that catalyze the respiration of pyruvic acid and other small organic molecules.

Here pyruvic acid is:

- Oxidized by NAD^+ producing NADH + H^+.

- Decarboxylated producing a molecule of:

  ○ Carbon dioxide (CO_2) and

  ○ A 2-carbon fragment of acetate bound to coenzyme A forming acetyl-CoA.

Citric Acid Cycle

The citric acid cycle is also known as the Krebs cycle (after the biochemist who worked it out) and the tricarboxylic acid cycle (TCA). The steps:

- The acetyl-CoA is donated to a molecule of oxaloacetic acid.

- The resulting molecule of citric acid (which gives its name to the process) undergoes the series of enzymatic steps shown in the diagram.

- The final step regenerates a molecule of oxaloacetic acid and the cycle is ready to turn again.

- Each of the 3 carbon atoms present in the pyruvate that entered the mitochondrion leaves as a molecule of carbon dioxide (CO_2).

- At 4 steps, a pair of electrons ($2e^-$) is removed and transferred to NAD^+ reducing it to $NADH^+ H^+$.

- At one step, a pair of electrons is removed from succinic acid and reduces the prosthetic group flavin adenine dinucleotide (FAD) to $FADH_2$.

The electrons of NADH and $FADH_2$ are transferred to the electron transport chain.

Electron Transport Chain

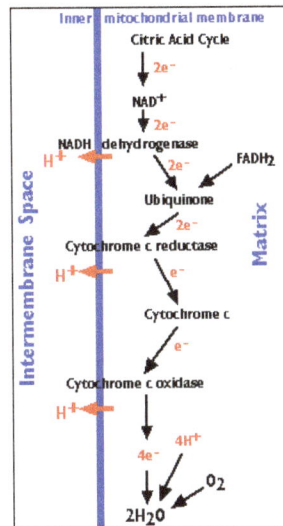

The electron transport chain consists of 3 complexes of integral membrane proteins:

- The NADH dehydrogenase,

- The cytochrome c reductase,

- The cytochrome c oxidase.

And two freely-diffusible molecules:

- Ubiquinone (also known as Coenzyme Q)

- Cytochrome c

That shuttle electrons from one complex to the next.

The electron transport chain accomplishes:

- The stepwise transfer of electrons from NADH (and $FADH_2$) to oxygen molecules to form (with the aid of protons) water molecules (H_2O).

- (Electrons from $FADH_2$ enter the electron transport chain through another integral membrane protein — complex II — reducing ubiquinone. (Cytochrome c can only transfer one electron at a time, so cytochrome c oxidase must wait until it has accumulated 4 of them before it can react with oxygen).

- Harnessing the energy released by this transfer to the pumping of protons (H^+) from the matrix to the intermembrane space.

- Approximately 20 protons are pumped into the intermembrane space as the 4 electrons needed to reduce oxygen to water pass through the respiratory chain.

- The gradient of protons formed across the inner membrane by this process of active transport forms a miniature battery.

- The protons can flow back down this gradient only by reentering the matrix through ATP synthase, another complex (complex V) of 16 integral membrane proteins in the inner membrane. The process is called chemiosmosis.

Chemiosmosis in Mitochondria

The energy released as electrons pass down the gradient from NADH to oxygen is harnessed by three enzyme complexes of the respiratory chain (I, III, and IV) to pump protons (H^+) against their concentration gradient from the matrix of the mitochondrion into the intermembrane space (an example of active transport).

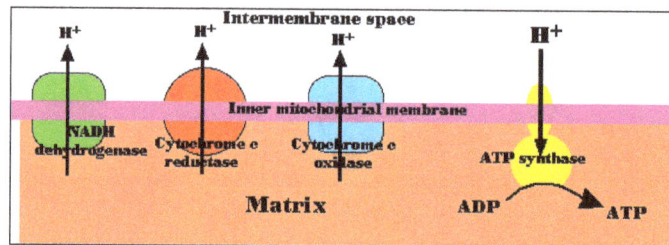

As their concentration increases there (which is the same as saying that the pH decreases), a strong diffusion gradient is set up. The only exit for these protons is through the ATP synthase complex. As in chloroplasts, the energy released as these protons flow down their gradient is harnessed to the synthesis of ATP. The process is called chemiosmosis and is an example of facilitated diffusion.

One-half of the 1997 Nobel Prize in Chemistry was awarded to Paul D. Boyer and John E. Walker for their discovery of how ATP synthase works.

ATPs

It is tempting to try to view the synthesis of ATP as a simple matter of stoichiometry (the fixed ratios of reactants to products in a chemical reaction). But (with 3 exceptions) it is not.

Most of the ATP is generated by the proton gradient that develops across the inner mitochondrial membrane. The number of protons pumped out as electrons drop from NADH through the respiratory chain to oxygen is theoretically large enough to generate, as they return through ATP synthase, 3 ATPs per electron pair (but only 2 ATPs for each pair donated by $FADH_2$).

With 12 pairs of electrons removed from each glucose molecule:

- 10 by NAD^+ (so 10x3=30);

- 2 by $FADH_2$ (so 2x2=4).

This could generate 34 ATPs.

Add to this the 4 ATPs that are generated by the 3 exceptions and one arrives at 38.

But:

- The energy stored in the proton gradient is also used for the active transport of several molecules and ions through the inner mitochondrial membrane into the matrix.

- NADH is also used as reducing agent for many cellular reactions.

So the actual yield of ATP as mitochondria respire varies with conditions. It probably seldom exceeds 30.

The Three Exceptions

A stoichiometric production of ATP does occur at:

- One step in the citric acid cycle yielding 2 ATPs for each glucose molecule. This step is the conversion of alpha-ketoglutaric acid to succinic acid.

- At two steps in glycolysis yielding 2 ATPs for each glucose molecule.

Mitochondrial DNA (mtDNA)

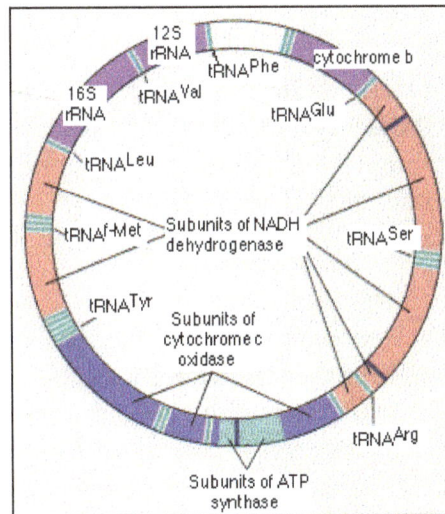

The human mitochondrion contains 5–10 identical, circular molecules of DNA. Each consists of 16,569 base pairs carrying the information for 37 genes which encode:

- 2 different molecules of ribosomal RNA (rRNA),

- 22 different molecules of transfer RNA (tRNA) (at least one for each amino acid),

- 13 proteins.

The rRNA and tRNA molecules are used in the machinery that synthesizes the 13 proteins.

The 13 proteins participate in building several protein complexes embedded in the inner mitochondrial membrane.

- 7 subunits that make up the mitochondrial NADH dehydrogenase,

- cytochrome b, a subunit of cytochrome c reductase,

- 3 subunits of cytochrome c oxidase,

- 2 subunits of ATP synthase.

Each of these protein complexes also requires subunits that are encoded by nuclear genes, synthesized in the cytosol, and imported from the cytosol into the mitochondrion. Nuclear genes also encode ~1,000 other proteins that must be imported into the mitochondrion.

Mutations in mtDNA cause Human Diseases

Mutations in 12 of the 13 protein-encoding mitochondrial genes have been found to cause human disease. Although many different organs may be affected, disorders of the muscles and brain are the most common. Perhaps this reflects the great demand for energy of both these organs. (Although representing only ~2% of our body weight, the brain consumes ~20% of the energy produced when we are at rest.)

Some of these disorders are inherited in the germline. In the vast majority of cases, the mutant gene is received from the mother because only very rarely do the mitochondria in sperm survive in the fertilized egg.

Other disorders are somatic; that is, the mutation occurs in the somatic tissues of the individual. These disorders can be caused not only by mutations in mtDNA, but also by mutations in the 228 nuclear genes that have also been implicated in human mitochondrial diseases. These latter mutations can be inherited from the father as well as the mother.

Exercise Intolerance

A number of humans who suffer from easily-fatigued muscles turn out to have a mutations in their cytochrome b gene. Curiously, only the mitochondria in their muscles have the mutation; the mtDNA of their other tissues is normal. Presumably, very early in their embryonic development, a mutation occurred in a cytochrome b gene in the mitochondrion of a cell destined to produce their muscles.

The severity of mitochondrial diseases varies greatly. The reason for this is probably the extensive mixing of mutant DNA and normal DNA in the mitochondria as they fuse with one another. A mixture of both is called heteroplasmy. The higher the ratio of mutant to normal, the greater the severity of the disease. In fact by chance alone, cells can on occasion end up with all their mitochondria carrying all-mutant genomes — a condition called homoplasmy (a phenomenon resembling genetic drift).

Mitochondrial Replacement Techniques

Only mothers can pass mutant mtDNA on to their offspring. Two techniques are under intense investigation, either of which could enable a mother to have children free of defective mitochondria.

Mutations in some 228 nuclear genes have also been implicated in human mitochondrial diseases, but mitochondrial replacement techniques will not be able to help with these.

Mitochondria have their own Genome

Many of the features of the mitochondrial genetic system resemble those found in bacteria. This has strengthened the theory that mitochondria are the evolutionary descendants of a bacterium that established an endosymbiotic relationship with the ancestors of eukaryotic cells early in the history of life on earth. However, many of the genes needed for mitochondrial function have since moved to the nuclear genome.

The recent sequencing of the complete genome of the endosymbiotic alpha-proteobacterium *Rickettsia prowazekii* has revealed a number of genes closely related to those found in mitochondria. This suggests a shared ancestry.

Aerobic Respiration

Aerobic respiration is the process by which oxygen-breathing creatures turn fuel, such as fats and sugars, into energy. Respiration is a process used by all cells to turn fuel, which contains stored energy, into a usable form. The product of respiration is a molecule called ATP, which can easily use the energy stored in its phosphate bonds to power chemical reactions the cell needs to survive.

Aerobic respiration is respiration that uses oxygen as a reactant. Aerobic respiration is much more efficient, and produces ATP much more quickly, than anaerobic respiration (respiration without oxygen). This is because oxygen is an excellent electron acceptor for the chemical reaction. Here, we will break down the process into simpler steps to illustrate how cellular respiration turns energy from glucose into a form that the cell can use to power its life functions.

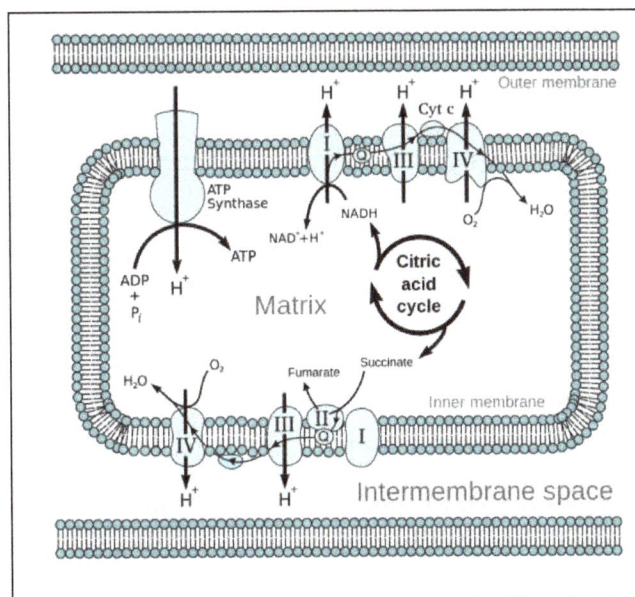

Common Steps between Aerobic Respiration and Anaerobic Respiration

Both aerobic respiration and anaerobic respiration use an electron transport chain to move energy from its long-term storage in sugars to a more usable form.

In respiration, the energy from sugar is moved into ATP, which can be used to power many chemical reactions necessary to a cell's survival.

Both aerobic and anaerobic respiration starts with the process of glycolysis. "Glycolysis," which literally means "sugar splitting," breaks a sugar molecule down into two smaller molecules.

In the process of glycolysis, two ATP molecules are consumed and four are produced. This result in a net gain of two ATP molecules produced for every sugar molecule broken down through glycolysis.

In cells that use oxygen, a sugar molecule is broken down into two molecules of pyruvate. In cells that do not have oxygen, the sugar molecule is broken down into other forms, such as lactate.

Although our cells normally use oxygen for respiration, which is much more efficient than anaerobic respiration, when we use ATP faster than we are getting oxygen molecules to our cells, our cells can perform anaerobic respiration to supply their needs for a few minutes.

Differences between Aerobic Respiration and Anaerobic Respiration

After glycolysis, different respiration chemistries take a few different paths:

- Cells that are deprived of oxygen but are not made for anaerobic respiration, like our own muscle cells, may leave the end products of glycolysis sitting around, obtaining only two ATP per sugar molecule they split.

- Cells that are made for anaerobic respiration may continue the electron transfer chain to extract more energy from the end products of glycolysis.

- Cells using aerobic respiration continue their electron transfer chain in a highly efficient process that ends up yielding 38 molecules of ATP from every sugar molecule.

After glycolysis, cells that do not use oxygen may use a different electron acceptor, such as sulfate or nitrate, to drive their reaction forward. These processes are called "fermentation." Some types of fermentation reactions actually have alcohol as their end product. So now you know where alcoholic drinks come from: the respiration processes of yeasts splitting sugars to produce energy.

Aerobic respiration, on the other hand, sends the pyruvate left over from glycolysis down a very different chemical path.

Aerobic Respiration and Weight Loss

Aerobic respiration is the process by which many cells, including our own, produce energy using food and oxygen. It also gives rise to carbon dioxide, which our bodies must then get rid of.

This equation explains why we need both food and oxygen, as both are reacted together to produce the ATP that allows our cells to function.

This equation also explains why we breathe out carbon dioxide – and how we lose weight. We breathe in O_2 and we breathe out the same number of molecules of CO_2. Where did the carbon atom come from? It comes from the food, such as sugar and fat that you've eaten.

This is also why you breathe harder and faster while performing calorie-burning activities: your body is using both oxygen and food at a faster-than-normal rate, and is producing more ATP to power your cells, along with more CO_2 waste product, as a result.

Unfortunately, simply breathing faster doesn't mean you'll unload more carbon: to lose carbon faster, your cells need to be consuming energy at a faster-than-normal rate. So get out those running shoes.

Aerobic Respiration Equation

The equation for aerobic respiration describes the reactants and products of all of its steps, including glycolysis. That equation is:

$$1 \text{ GLUCOSE} + 6O_2 \rightarrow 6CO_2 + 6 \text{ H}_2\text{O} + 38 \text{ ATP}$$

The reactions of aerobic respiration can be broken down into four stages, described below:

Steps of Aerobic Respiration

1. Glycolysis: In aerobic cells, the equation for glycolysis is:

$$\text{GLUCOSE} + 2 \text{ HPO}_4^{2-} + 2 \text{ ADP}^{3-} + 2 \text{ NAD}^+ \rightarrow 2 \text{ PYRUVATE}^- + 2 \text{ ATP}^{4-} + 2 \text{ NADH} + 2 \text{ H}^+ + 2 \text{ H}_2\text{O}$$

Glycolysis in aerobic respiration refers to the splitting of a sugar molecule into two pyruvate molecules. This process creates two ATP molecules.

We will notice that this process also creates NADH from NAD^+. This is important because later in the process of cellular respiration, NADH will power the formation of much more ATP through the mitochondria's electron transport chain.

Pyruvate is then processed to turn it into fuel for the citric acid cycle, using the process of oxidative decarboxylation.

2. Oxidative decarboxylation of pyruvate:

$$2 \text{ (PYRUVATE}^- + \text{COENZYME A} + \text{NAD}^+ \rightarrow \text{ACETYL COA} + \text{CO}_2 + \text{NADH)}$$

In this process, pyruvate is combined with coenzyme A to produce acetyl-CoA. We will note that more NADH is created in this step. This means more fuel to create more ATP later in the process of cellular respiration. This is important because acetyl-CoA is an ideal fuel for the citric acid cycle, which can in turn power the process of oxidative phosphorylation in the mitochondria, which produces huge amounts of ATP.

3. Citric acid cycle:

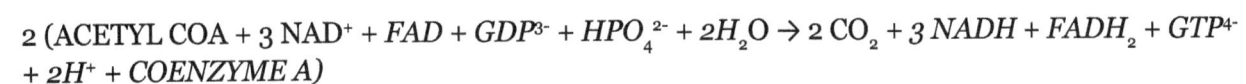

$$2 \text{ (ACETYL COA} + 3 \text{ NAD}^+ + \text{FAD} + \text{GDP}^{3-} + \text{HPO}_4^{2-} + 2\text{H}_2\text{O} \rightarrow 2 \text{ CO}_2 + 3 \text{ NADH} + \text{FADH}_2 + \text{GTP}^{4-} + 2\text{H}^+ + \text{COENZYME A)}$$

In the citric acid cycle, both NADH and $FADH_2$ – another carrier of electrons for the electron transport chain – are created. All the NADH and $FADH_2$ created in the preceding steps now come into play in the process of oxidative phosphorylation.

4. Oxidative phosphorylation:

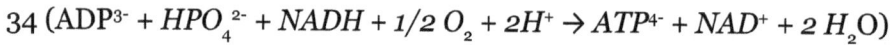

$$34 \; (ADP^{3-} + HPO_4{}^{2-} + NADH + 1/2 \; O_2 + 2H^+ \rightarrow ATP^{4-} + NAD^+ + 2 \; H_2O)$$

Oxidative phosphorylation uses the folded membranes within the cell's mitochondria to produce huge amounts of ATP. In this process, NADH and $FADH_2$ donate the electrons they obtained from glucose during the previous steps of cellular respiration to the electron transport chain in the mitochondria's membrane. The electron transport chain consists of a number of complexes in the mitochondrial membrane, including complex I, Q, complex III, cytochrome C, and complex IV.

All of these ultimately serve to pass electrons from higher to lower energy levels, harvesting bits of their energy in the process. This energy is used to power proton pumps, which in turn power ATP formation. Just like the sodium-potassium pump of the cell membrane, the proton pumps of the mitochondrial membrane are used to create a concentration gradient which can be used to power other processes. In the case of the mitochondria's proton gradient, the protons that are transported across the membrane using the energy harvested from NADH and $FADH_2$ "want" to pass through channel proteins from their area of high concentration to their area of low concentration.

These channel proteins are actually ATP synthase – the enzyme that makes ATP. When protons pass through ATP synthase, they drive the formation of ATP. This process is why mitochondria are referred to as "the powerhouses of the cell." The mitochondria's electron transport chain makes nearly 90% of all the ATP produced by the cell from breaking down food.

This is also the process that requires oxygen. Without oxygen molecules to accept the depleted electrons at the end of the electron transport chain, the electrons would back up and the process of ATP creation would not be able to continue. No wonder we need oxygen to live.

Function of Aerobic Respiration

Aerobic respiration produces ATP, which is then used to power other life-sustaining functions, such as the action of the sodium-potassium pump, which allows us to move, think, and perceive the world around us; the actions of many enzymes; and the actions of countless other proteins that sustain life.

Cell Reproduction

Cell reproduction is the process by which cells divide to form new cells. Each time a cell divides, it makes a copy of all of its chromosomes, which are tightly coiled strands of DNA, the genetic material that holds the instructions for all life, and sends an identical copy to the new cell that is created.

Humans have 46 Chromosomes within each of their body cells. Other species have different numbers of Chromosomes, however. One species of fern has 1262 of them. As you might guess, the number of chromosomes does not directly impact the complexity of an organism. As chromosomes vary in size, one human chromosome can hold genetic information equivalent to the amound ot genetic information in many chromosomes from another organism.

A chromosomes consists of two halves, called Chromatids. These halves are divided in their center by a centromere. This structure is what attaches to spindle fibers during mitosis to pull one chromatid to each side of the cell when it divides.

In humans, 44 of the chromosomes consist of autosomes, and the remaining two are the sex chromosomes. These chromosomes determine the gender of the organism. (A male has an X and a Y, while a female has to Xs).

In addition, all the chromosomes in an organism excluding the sex chromosomes are part of a *homologous pair*. They contain genes to control the same traits but the genes do not have the same instructions. For example, one chromosome might have the genes for brown eyes while its homolouge might have genes for blue eyes. One homolouge is inherited from the mother while the other is inherited from the father.

The Cell Cycle

The cell cycle is the of steps that cells take to grow, develop, and reproduce. It can be broken down into five steps:

1. G1 Phase
2. S Phase
3. G2 Phase
4. M Phase
5. Cytokinesis.

G1 Phase

During the G1 Phase, the cell grows and stores up energy that it will use during cell division. Nutrients are taken in and all the usual cell processes take place. Once cells are fully grown, they proceed on to the S Phase.

S Phase

During the S Phase, the DNA in the cell's nucleus is copied. This means that the cell then attains two copies of the entire necessary DNA for normal cell activity, leaving a full set to be transferred into the new cell that will be created after the cell divides.

G2 Phase

During this phase, the cell prepares for cell division. This phase represents a time gap between the time when the cell copies its DNA and when it divides.

M Phase

During this phase, cell division takes place through Mitosis.

Cytokinesis

During Cytokinesis, the cytoplasm in the cell divides and the cell's membrane pinches inward and the cell begins to divide. Also, when plant cells divide, a cell plate forms between the two new cells to divide them. After this step, the new cell and sometimes the original cell also restart the cell cycle by beginning G1 Phase again. However, sometimes cells enter G0 phase, which is a phase where cells exit the cell cycle after they are fully grown and continue to serve their purpose in an organism.

Other Methods of Cell Reprocuction

Several other methods of cell reproduction exist. These include meiosis and binary fission. During binary fission, bacterial cells divide asexually.

Cell Division

Cell division is the process cells go through to divide. There are several types of cell division, depending upon what type of organism is dividing. Organisms have evolved over time to have different and more complex forms of cell division. Most prokaryotes, or bacteria, use binary fission to divide the cell. Eukaryotes of all sizes use *mitosis* to divide. Sexually-reproducing eukaryotes use a special form of cell division called *meiosis* to reduce the genetic content in the cell. This is necessary in sexual reproduction because each parent must give only half of the required genetic material, otherwise the offspring would have too much DNA, which can be a problem.

Types of Cell Division

Prokaryotic Cell Division

Prokaryotes replicate through a type of cell division known as *binary fission*. Prokaryotes are simple organism, with only one membrane and no division internally. Thus, when a prokaryote divides, it simply replicates the DNA and splits in half. The process is a little more complicated than this, as DNA must first be unwound by special proteins. Although the DNA in prokaryotes usually exists in a ring, it can get quite tangled when it is being used by the cell. To copy the DNA efficiently, it must be stretched out. This also allows the two new rings of DNA created to be separated after they are produced. The two strands of DNA separate into two different sides of the prokaryote cell. The cell then gets longer, and divides in the middle.

The DNA is the tangled line. The other components are labeled. Plasmids are small rings of DNA that also get copied during binary fission and can be picked up in the environment, from dead cells that break apart. These plasmids can then be further replicated. If a plasmidis beneficial, it will increase in a population. This is in part how antibiotic resistance in bacteria happens. The ribosomes are small protein structures that help produce proteins. They are also replicated so each cell can have enough to function.

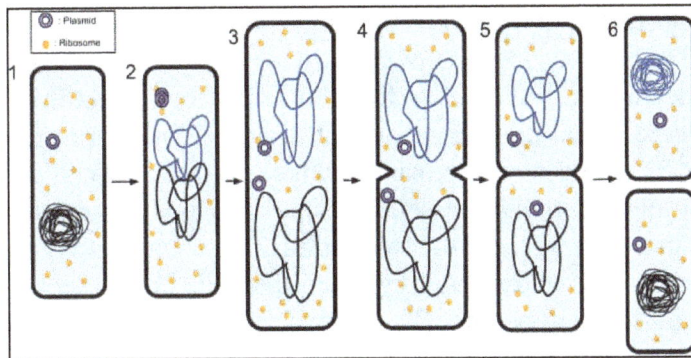

Eukaryotic Cell Division: Mitosis

Eukaryotic organisms have membrane bound organelles and DNA that exists on chromosomes, which makes cell division harder. Eukaryotes must replicate their DNA, organelles, and cell mechanisms before dividing. Many of the organelles divide using a process that is essentially *binary fission*, leading scientist to believe that eukaryotes were formed by prokaryotes living inside of other prokaryotes.

After the DNA and organelles are replicated during *interphase* of the cell cycle, the eukaryote can begin the process of mitosis. The process begins during prophase, when the chromosomes condense. If mitosis proceeded without the chromosomes condensing, the DNA would become tangled and break. Eukaryotic DNA is associated with many proteins which can fold it into complex structures. As mitosis proceeds to *metaphase* the chromosomes are lined up in the middle of the cell. Each half of a chromosome, known as *sister chromatids* because they are replicated copies of each other, gets separated into each half of the cell as mitosis proceeds. At the end of mitosis, another process called *cytokinesis* divides the cell into two new daughter cells.

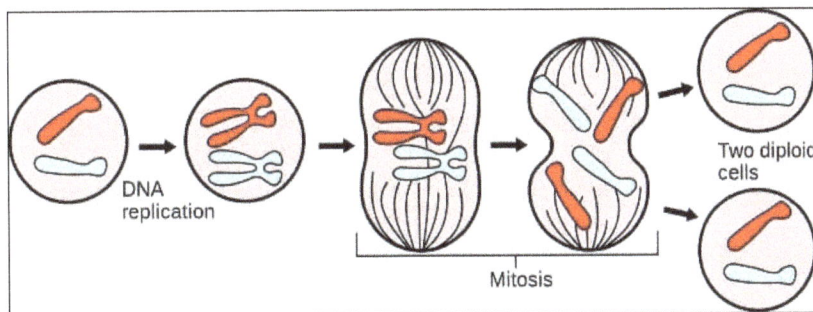

All eukaryotic organisms use mitosis to divide their cells. However, only single-celled organisms use mitosis as a form of reproduction. Most multicellular organisms are sexually reproducing and combine their DNA with that of another organism to reproduce. In these cases, organisms need a different method of cell division. Mitosis yields identical cells, but meiosis produces cells with half the genetic information of a regular cell, allowing two cells from different organisms of the same species to combine.

Eukaryotic Cell Division: Meiosis

In sexually reproducing animals, it is usually necessary to reduce the genetic information before fertilization. Some plants can exist with too many copies of the genetic code, but in most organisms it is highly detrimental to have too many copies. Humans with even one extra copy of one chromosome can experience detrimental changes to their body. To counteract this, sexually reproducing organisms undergo a type of cell division known as *meiosis*. As before mitosis, the DNA and organelles are replicated. The process of meiosis contains two different cell divisions, which happen back-to-back. The first meiosis, *meiosis I*, separates homologous chromosomes. The homologous chromosomes present in a cell represent the two alleles of each gene an organism has. These alleles are recombined and separated, so the resulting daughter cells have only one allele for each gene, and no homologous pairs of chromosomes. The second division, *meiosis II*, separated the two copies of DNA, much like in mitosis. The end result of meiosis in one cell is 4 cells, each with only one copy of the genome, which is half the normal number.

Organisms typically package these cells into *gametes*, which can travel into the environment to find other gametes. When two gametes of the right type meet, one will fertilize the other and produce a *zygote*. The zygote is a single cell that will undergo mitosis to produce the millions of cells necessary for a large organism. Thus, most eukaryotes use both mitosis and meiosis, but at different stages of their lifecycle.

Cell Division Stages

Depending upon which type of cell division an organism uses, the stages can be slightly different.

Mitosis Stages

Mitosis starts with *prophase* in which the chromosome is condensed. The cell proceeds to *metaphase* where the chromosomes are aligned on the metaphase plate. Then the chromosomes are separated in *anaphase* and the cell's cytoplasm is pinched apart during *telophase*. *Cytokinesis* is the final process that breaks the cell membrane and divides the cell into two.

Meiosis Stages

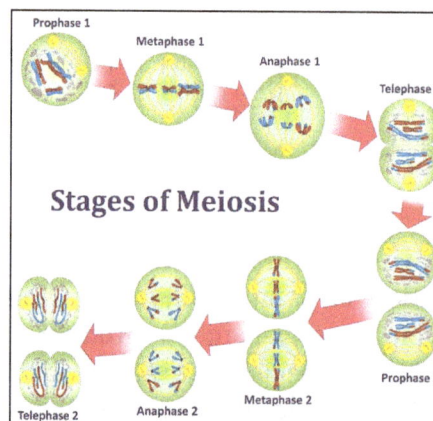

The stages of meiosis are similar to mitosis, but the chromosomes act differently. Meiosis has two phases, which include two separate cell divisions without the DNA replicating between them. *Meiosis I* and *meiosis II* have the same 4 stages as mitosis: prophase, metaphase, anaphase, and telophase. Cytokinesis concludes both rounds of meiosis.

In prophase I, the chromosomes are condensed. In metaphase I, the chromosomes line up across from their homologous pairs. When they are separated in anaphase I and telophase I, there is only one form of each gene in each cell, known as a reduction division. Meiosis II precedes in the same manner as mitosis, which sister chromatids dividing on the metaphase plate. By telophase II, there are 4 cells, each with half of the alleles as the parent cell and only a single copy of the genome. The cells can now become gametes and fuse together to create new organisms.

Binary Fission

Binary fission is a form of asexual reproduction used by members of domains archaea and bacteria among other organisms. Like mitosis (in eukaryotic cells), it results in cell division of the original cell to produce two viable cells that can repeat the process.

Though the concept of binary fission is similar to mitosis, there are a few major differences between the two. Whereas binary fission is a method of propagation used by bacteria and archaea, mitosis occurs in eukaryotic cells.

Moreover, the two occur for different reasons in cells. However, the two processes go through a number of phases involving DNA division followed by splitting of the cell into two daughter cells. Some eukaryotes like paramecium and amoebae can use binary fission as a means of propagation. For simple organisms like bacteria, cell division (for propagation) is dependent on a form of asexual reproduction known as binary fission. Compared to the cell structure of eukaryotic organisms, the cell structure of prokaryotes (e.g. bacteria and archaea) is very simple. For this reason, they can only reproduce asexually through binary fission, a relatively simple method of reproduction.

During cell division, the DNA molecule of prokaryotes (a single circular chromosome) is first uncoiled before being replicated to produce two chromosomes. The two molecules then start moving to the opposite poles of the cells as the cell pulls apart. This results in the cell increasing in size (length) before ultimately splitting into two.

Before the cell divides into two, cell organelles (e.g. plasmids, ribosome etc) increase in number thus allowing each of the daughter cells to contain approximately equal number of organelles (as well as cytoplasm).

Binary fission in prokaryotes can be divided into four main phases that include:

- Chromosome replication: The single, circular chromosome is uncoiled and copied to form a new chromosome thus doubling the genetic content.

- Cell growth: Following chromosome replication, the cell grows and increases in size in preparation for binary fission. This growth is accompanied by an increase in the volume of the cytoplasm with some of the organelles increasing in number. This phase is also characterized by the two strands starting to migrate to opposite poles of the cells.

- Chromosome segregation: In this stage, the cell elongates as a septum forms at the middle (transversely). It is also at this point that the chromosomes completely separate.

- Cell splitting: In this stage, a new cell wall is formed. Ultimately, the cell splits along the middle (at the septum) dividing the cell into two new daughter cells, each of which contains nuclear material and other cell organelles.

Organelles such as mitochondria also use this method of cell division (binary fission) to increase in number.

Movement of chromosome strands to the opposite poles of the cell requires energy.

In any of the 4 stages, issues may arise resulting in various abnormalities.

There are four types of binary fission that include:

- Irregular binary fission,

- Longitudinal binary fission,

- Transverse binary fission,

- Oblique binary fission.

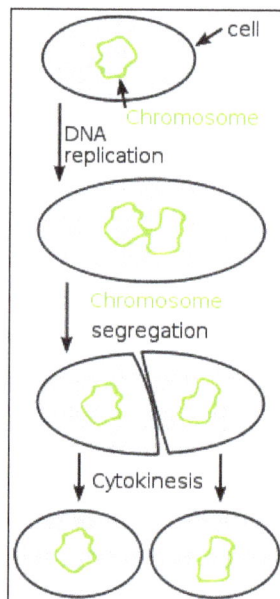

Binary Fission in Single-celled Organisms

Binary Fission in Paramecium (Asexual Reproduction)

Paramecium is a genus consisting of such ciliates as Paramecium aurelia, Paramecium bursaria and Paramecium caudatum among many others.

Paramecia (members of the genus paramecium) are shaped like a slipper or sole of a shoe and covered by cilia that allow them to move from place to another and feed. Unlike Bacteria, paramecium consists of eukaryotic cells which means that they have well-organized cells. Although Paramecia can reproduce sexually, asexual reproduction (binary fission) is the primary means of reproduction that occurs in favorable conditions.

For Paramecium, asexual reproduction is in the form of transverse binary fission. While the organism has two nuclei (a large macronucleus and a small micronucleus) it is the small nucleus that is involved/responsible for reproduction.

In favorable environmental conditions (water etc.) the ciliate stops feeding as the oral groove disappears. This is then followed by the micronucleus dividing into two and the two micronuclei moving to the pole ends of the cell.

Once the small nucleus has divided, the macronucleus also divides into two and the two move to either end of the cell. Lastly, cytokinesis occurs along the transverse axis to produce two similar cells (This involves the cytoplasm dividing at the right angle to the long axis of the cell).

The micronucleus of paramecium divides through mitosis while the larger nucleus divides through amitosis. Here, the genetic material in the nucleus (micronucleus) divides through mitosis. The nucleus goes through four stages of mitosis that include prophase, metaphase, anaphase, and telophase.

During micronucleus division, all energy is used for separation and duplication. Therefore, cell growth does not occur at this stage.

While Paramecium also uses a complex method of sexual reproduction, binary fission occurs when the partners (conjugant) separate.

Binary Fission in Bacteria

Bacteria use several methods of asexual methods for reproduction. These include:

- Binary fission

- Zoogloea stage

- Formation of conidia and gonidia

- Budding

- Fragmentation.

Although different types of bacteria can use any of the aforementioned methods, binary fission is the most common method of asexual reproduction that occurs when environmental conditions are favorable (water, optimal temperature range etc.).

Binary fission in bacteria involves two main stages that include:

Genome Replication

For bacteria, as is the case with Paramecium, binary fission starts with replication of the genome (bacterial chromosome in the nucleoid). Here, replication enzymes copy the chromosome strand starting at the origin of replication then continues separating the strand into two. Replication produces two circular daughter cells as the cell elongates. The origins then move to either end of the cell as DNA is copied.

Septum Formation/Cell Division

During this phase, plasma membrane's peripheral ring invaginates dividing the cell into two. This is also accompanied by the formation of a double-membranous septum before the two cells completely separate into two similar cells.

Unlike Paramecium, division of the nucleoid does not involve mitosis.

Under favorable environmental conditions, binary fission may take about 30 minutes.

Binary fission is susceptible to a wide range of environmental changes. For this reason, other forms of asexual reproduction come to play during unfavorable environment conditions.

Binary Fission in Amoeba

Amoeba is a genus consisting of such eukaryotic organisms as Amoeba proteus. Amoebae lack a definite shape and move through temporary projections known as pseudopodia (false feet).

Like many other eukaryotes, the cytoplasm and cellular contents of Amoebae are contained in a cell membrane while DNA is contained in the nucleus.

For such species as Amoeba proteus, sexual reproduction is achieved through binary fission (a form of asexual reproduction). However, it may also involve multiple fission or sporulation.

As is the case with Paramecium, also a eukaryote, genetic material is replicated through mitosis. Here, the DNA strand goes through the four stages of mitosis (prophase, metaphase, anaphase and telophase).

By the end of telophase, two daughter nuclei are formed with a lattice forming beneath each nuclear membrane. Nucleus division is followed by cytokinesis which divides the cytoplasm and ultimately produces two daughter cells that are almost identical in terms of cellular content.

Unlike bacteria, multipolar nuclear spindle are produced during binary fission in amoeba.

Reproduction in amoeba occurs through irregular binary fission. Here, cytokinesis occurs along a plane that is perpendicular to the plane of karyokinesis.

Differences between Binary Fission and Mitosis

Apart from the fact that binary fission is common in prokaryotes and mitosis in eukaryotes, some of the other major differences between the two methods of cell division include:

- Whereas spindle apparatus are produced during mitosis (to separate chromosomes) they are not formed in binary fission.

- Eukaryotes have cell organelles that double in interphase in preparation of mitosis. However, given that prokaryotes are simpler cells, only ribosome and a few other cell components increase in number before binary fission starts.

- Whereas binary fission is only used as a means of reproduction by prokaryotes, mitosis is used for several functions that include cell replacement, growth and development of the organism.

- Due to complexity of eukaryotic cells, mitosis takes longer than binary fission - Binary fission occurs rapidly.

Meiosis and Sexual Reproduction

Meiosis

Sexual reproduction occurs only in eukaryotes. During the formation of gametes, the number of chromosomes is reduced by half, and returned to the full amount when the two gametes fuse during fertilization.

Ploidy

Haploid and diploid are terms referring to the number of sets of chromosomes in a cell. Gregor Mendel determined his peas had two sets of alleles, one from each parent. Diploid organisms are those with two (di) sets. Human beings (except for their gametes), most animals and many plants are diploid. We abbreviate diploid as 2n. Ploidy is a term referring to the number of sets of chromosomes. Haploid organisms/cells have only one set of chromosomes, abbreviated as n. Organisms with more than two sets of chromosomes are termed polyploid. Chromosomes that carry the same genes are termed homologous chromosomes. The alleles on homologous chromosomes may differ, as in the case of heterozygous individuals. Organisms (normally) receive one set of homologous chromosomes from each parent.

Meiosis is a special type of nuclear division which segregates one copy of each homologous chromosome into each new "gamete". Mitosis maintains the cell's original ploidy level (for example, one diploid 2n cell producing two diploid 2n cells; one haploid n cell producing two haploid n cells; etc.). Meiosis, on the other hand, reduces the number of sets of chromosomes by half, so that when gametic recombination (fertilization) occurs the ploidy of the parents will be reestablished.

Most cells in the human body are produced by mitosis. These are the somatic (or vegetative) line cells. Cells that become gametes are referred to as germ line cells. The vast majority of cell divisions in the human body are mitotic, with meiosis being restricted to the gonads.

Life Cycles

Life cycles are a diagrammatic representation of the events in the organism's development and reproduction. When interpreting life cycles, pay close attention to the ploidy level of particular parts of the cycle and where in the life cycle meiosis occurs. For example, animal life cycles have a dominant diploid phase, with the gametic (haploid) phase being a relative few cells. Most of the cells in your body are diploid, germ line diploid cells will undergo meiosis to produce gametes, with fertilization closely following meiosis.

Plant life cycles have two sequential phases that are termed alternation of generations. The sporophyte phase is "diploid", and is that part of the life cycle in which meiosis occurs. However, many plant species are thought to arise by polyploidy, and the use of "diploid" in the last sentence was meant to indicate that the greater number of chromosome sets occur in this phase. The gametophyte phase is "haploid", and is the part of the life cycle in which gametes are produced (by mitosis of haploid cells). In flowering plants (angiosperms) the multicelled visible plant (leaf, stem, etc.)

is sporophyte, while pollen and ovaries contain the male and female gametophytes, respectively. Plant life cycles differ from animal ones by adding a phase (the haploid gametophyte) after meiosis and before the production of gametes.

Many protists and fungi have a haploid dominated life cycle. The dominant phase is haploid, while the diploid phase is only a few cells (often only the single celled zygote, as in *Chlamydomonas*). Many protists reproduce by mitosis until their environment deteriorates, then they undergo sexual reproduction to produce a resting zygotic cyst.

Phases of Meiosis

Two successive nuclear divisions occur, Meiosis I (Reduction) and Meiosis II (Division). Meiosis produces 4 haploid cells. Mitosis produces 2 diploid cells. The old name for meiosis was reduction/division. Meiosis I reduces the ploidy level from 2n to n (reduction) while Meiosis II divides the remaining set of chromosomes in a mitosis-like process (division). Most of the differences between the processes occur during Meiosis I.

Prophase I

Prophase I has a unique event the pairing (by an as yet undiscovered mechanism) of homologous chromosomes. Synapsis is the process of linking of the replicated homologous chromosomes. The resulting chromosome is termed a tetrad, being composed of two chromatids from each chromosome, forming a thick (4-strand) structure. Crossing-over may occur at this point. During crossing-over chromatids break and may be reattached to a different homologous chromosome.

The alleles on this tetrad:

- A B C D E F G
- A B C D E F G
- a b c d e f g
- a b c d e f g

Will produce the following chromosomes if there is a crossing-over event between the 2nd and 3rd chromosomes from the top:

- A B C D E F G
- A B c d e f g
- a b C D E F G
- a b c d e f g

Thus, instead of producing only two types of chromosome (all capital or all lower case), four different chromosomes are produced. This doubles the variability of gamete genotypes. The occurrence of a crossing-over is indicated by a special structure, a chiasma (plural chiasmata) since the recombined inner alleles will align more with others of the same type (e.g. a with a, B with B). Near the end of Prophase I, the homologous chromosomes begin to separate slightly, although they remain attached at chiasmata.

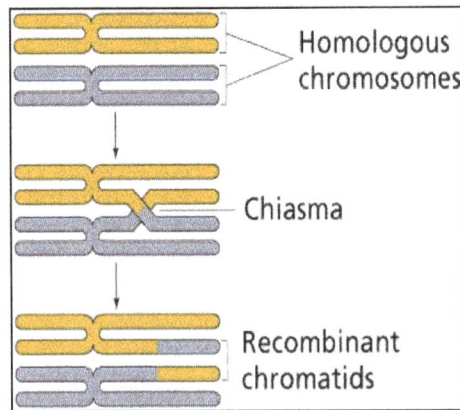

Crossing-over between homologous chromosomes produces chromosomes with new associations of genes and alleles.

Events of Prophase I (save for synapsis and crossing over) are similar to those in Prophase of mitosis: chromatin condenses into chromosomes, the nucleolus dissolves, nuclear membrane is disassembled, and the spindle apparatus forms.

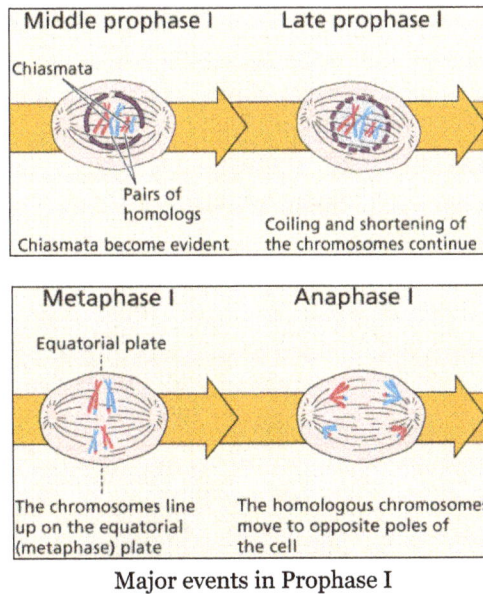

Major events in Prophase I

Metaphase

Metaphase I is when tetrads line-up along the equator of the spindle. Spindle fibers attach to the centromere region of each homologous chromosome pair. Other metaphase events as in mitosis.

Anaphase

Anaphase I is when the tetrads separate, and are drawn to opposite poles by the spindle fibers. The centromeres in Anaphase I remain intact.

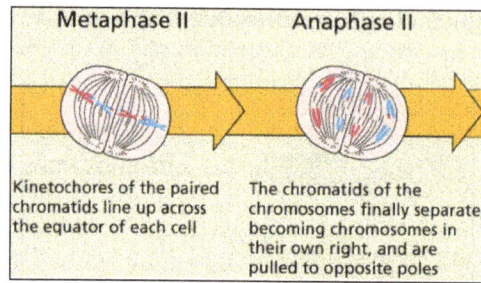

Metaphase II	Anaphase II

Kinetochores of the paired chromatids line up across the equator of each cell

The chromatids of the chromosomes finally separate, becoming chromosomes in their own right, and are pulled to opposite poles

Events in prophase and metaphse I

Telophase I

Telophase I is similar to Telophase of mitosis, except that only one set of (replicated) chromosomes is in each "cell". Depending on species, new nuclear envelopes may or may not form. Some animal cells may have division of the centrioles during this phase.

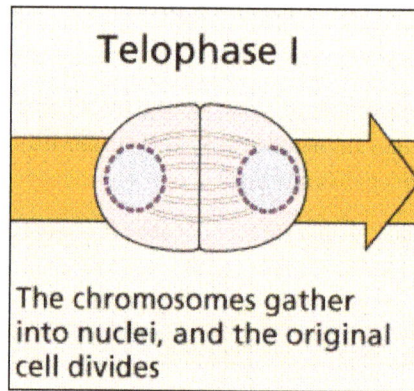

Telophase I

The chromosomes gather into nuclei, and the original cell divides

The events of Telophase I

Prophase II

During Prophase II, nuclear envelopes (if they formed during Telophase I) dissolve, and spindle fibers reform. All else is as in Prophase of mitosis. Indeed Meiosis II is very similar to mitosis.

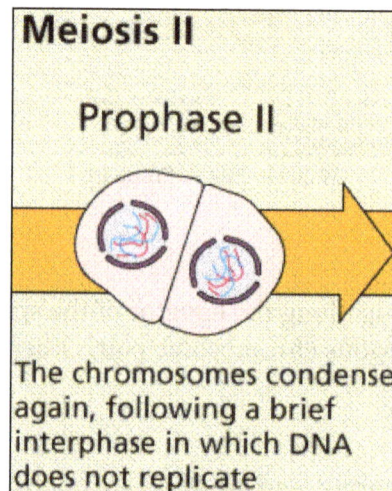

Meiosis II

Prophase II

The chromosomes condense again, following a brief interphase in which DNA does not replicate

The events of Prophase II

Metaphase II

Metaphase II is similar to mitosis, with spindles moving chromosomes into equatorial area and attaching to the opposite sides of the centromeres in the kinetochore region.

Anaphase II

During Anaphase II, the centromeres split and the former chromatids (now chromosomes) are segregated into opposite sides of the cell.

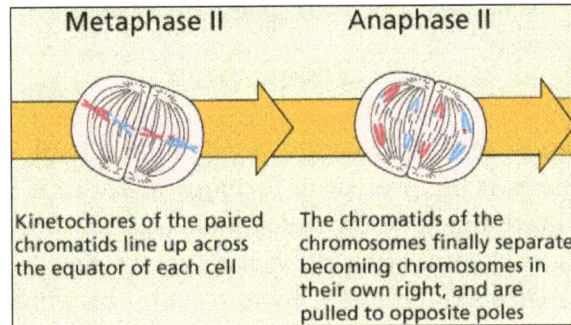

The events of Metaphase II and Anaphase II

Telophase II

Telophase II is identical to Telophase of mitosis. Cytokinesis separates the cells.

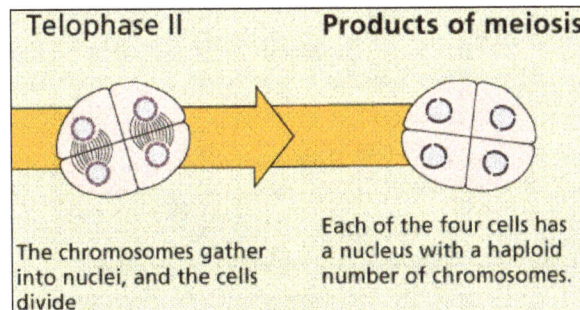

The events of Telophase II

Comparison of Mitosis and Meiosis

Mitosis maintains ploidy level, while meiosis reduces it. Meiosis may be considered a reduction phase followed by a slightly altered mitosis. Meiosis occurs in a relative few cells of a multicellular organism, while mitosis is more common.

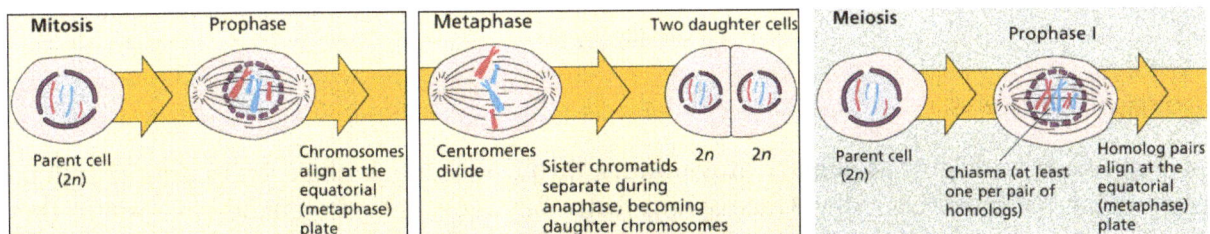

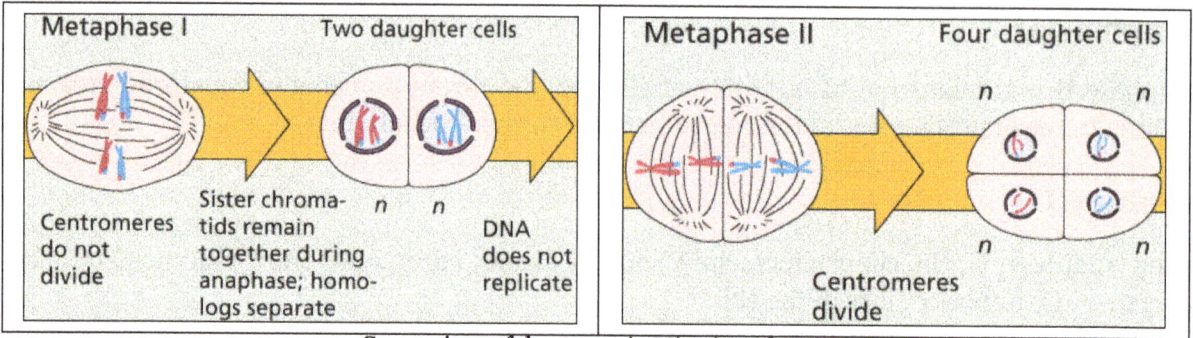

| Metaphase I | Two daughter cells | Metaphase II | Four daughter cells |

Centromeres do not divide   Sister chroma-tids remain together during anaphase; homo-logs separate   *n*   *n*   DNA does not replicate   Centromeres divide   *n*   *n*   *n*   *n*

Comparison of the events in Mitosis and Meiosis

Gametogenesis

Gametogenesis is the process of forming gametes (by definition haploid, n) from diploid cells of the germ line. Spermatogenesis is the process of forming sperm cells by meiosis (in animals, by mitosis in plants) in specialized organs known as gonads (in males these are termed testes). After division the cells undergo differentiation to become sperm cells. Oogenesis is the process of forming an ovum (egg) by meiosis (in animals, by mitosis in the gametophyte in plants) in specialized gonads known as ovaries. Whereas in spermatogenesis all 4 meiotic products develop into gametes, oogenesis places most of the cytoplasm into the large egg. The other cells, the polar bodies, do not develop. This all the cytoplasm and organelles go into the egg. Human males produce 200,000,000 sperm per day, while the female produces one egg (usually) each menstrual cycle.

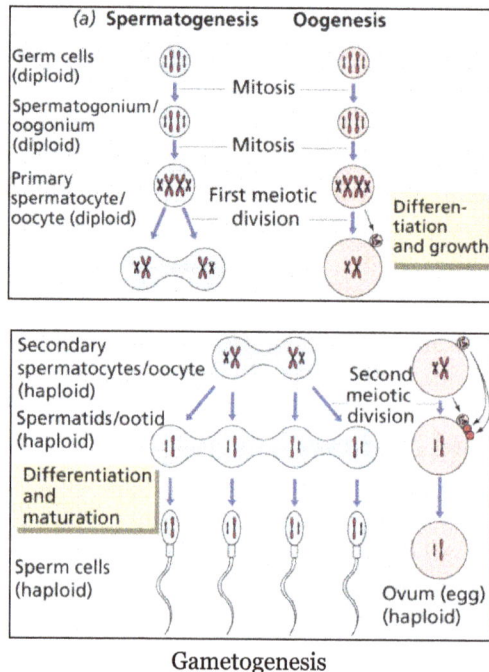

Gametogenesis

Spermatogenesis

Sperm production begins at puberty at continues throughout life, with several hundred million sperm being produced each day. Once sperm form they move into the epididymis, where they mature and are stored.

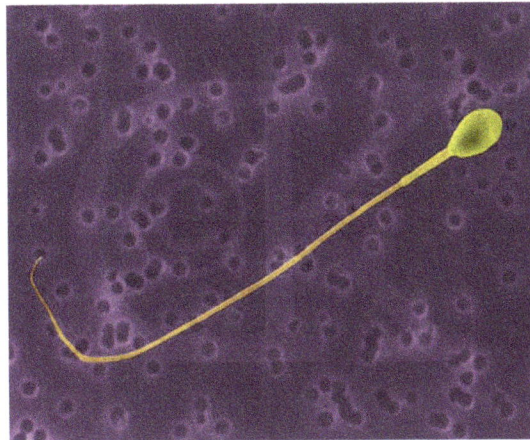

Human Sperm (SEM x5,785)

Oogenesis

The ovary contains many follicles composed of a developing egg surrounded by an outer layer of follicle cells. Each egg begins oogenesis as a primary oocyte. At birth each female carries a lifetime supply of developing oocytes, each of which is in Prophase I. A developing egg (secondary oocyte) is released each month from puberty until menopause, a total of 400-500 eggs.

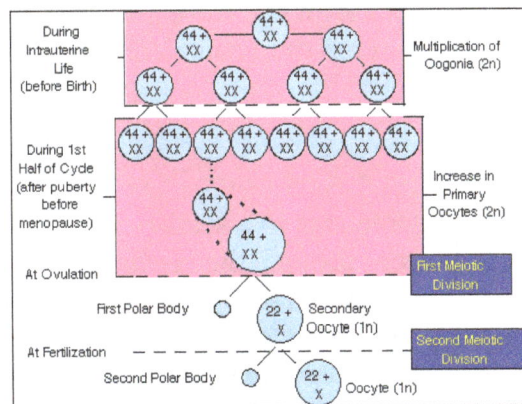

Oogenesis

Cell Cycle

The cell cycle is an ordered series of events involving cell growth and cell division that produces two new daughter cells. Cells on the path to cell division proceed through a series of precisely timed and carefully regulated stages of growth, DNA replication, and division that produces two identical (clone) cells. The cell cycle has two major phases: interphase and the mitotic phase. During interphase, the cell grows and DNA is replicated. During the mitotic phase, the replicated DNA and cytoplasmic contents are separated, and the cell divides.

The cell cycle consists of interphase and the mitotic phase. During interphase, the cell grows and the nuclear DNA is duplicated. Interphase is followed by the mitotic phase. During the mitotic phase, the duplicated chromosomes are segregated and distributed into daughter nuclei. The cytoplasm is usually divided as well, resulting in two daughter cells.

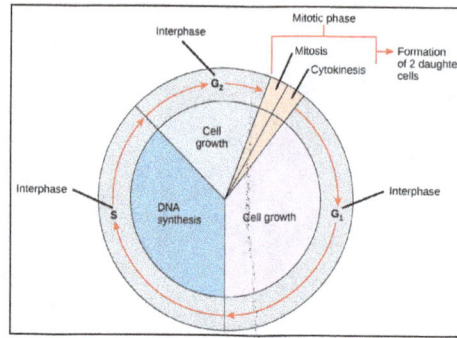

Interphase

During interphase, the cell undergoes normal growth processes while also preparing for cell division. In order for a cell to move from interphase into the mitotic phase, many internal and external conditions must be met. The three stages of interphase are called G_1, S, and G_2.

G_1 Phase

The first stage of interphase is called the G_1 phase (first gap) because, from a microscopic aspect, little change is visible. However, during the G_1 stage, the cell is quite active at the biochemical level. The cell is accumulating the building blocks of chromosomal DNA and the associated proteins as well as accumulating sufficient energy reserves to complete the task of replicating each chromosome in the nucleus.

S Phase (Synthesis of DNA)

Throughout interphase, nuclear DNA remains in a semi-condensed chromatin configuration. In the S phase, DNA replication can proceed through the mechanisms that result in the formation of identical pairs of DNA molecules—sister chromatids—that are firmly attached to the centromeric region. The centrosome is duplicated during the S phase. The two centrosomes will give rise to the mitotic spindle, the apparatus that orchestrates the movement of chromosomes during mitosis. At the center of each animal cell, the centrosomes of animal cells are associated with a pair of rod-like objects, the centrioles, which are at right angles to each other. Centrioles help organize cell division. Centrioles are not present in the centrosomes of other eukaryotic species, such as plants and most fungi.

G_2 Phase

In the second gap , the cell replenishes its energy stores and synthesizes proteins necessary for chromosome manipulation. Some cell organelles are duplicated, and the cytoskeleton is dismantled to provide resources for the mitotic phase. There may be additional cell growth during G_2. The final preparations for the mitotic phase must be completed before the cell is able to enter the first stage of mitosis.

Mitotic Phase

The mitotic phase is a multi-step process during which the duplicated chromosomes are aligned, separated, and move into two new, identical daughter cells. The first portion of the mitotic phase is called karyokinesis, or nuclear division. The second portion of the mitotic phase, called cytokinesis, is the physical separation of the cytoplasmic components into the two daughter cells.

Karyokinesis (Mitosis)

Karyokinesis, also known as mitosis, is divided into a series of phases—prophase, prometaphase, metaphase, anaphase, and telophase—that result in the division of the cell nucleus. Karyokinesis is also called mitosis.

During prophase, the "first phase," the nuclear envelope starts to dissociate into small vesicles, and the membranous organelles (such as the Golgi complex or Golgi apparatus, and endoplasmic reticulum), fragment and disperse toward the periphery of the cell. The nucleolus disappears (disperses). The centrosomes begin to move to opposite poles of the cell. Microtubules that will form the mitotic spindle extend between the centrosomes, pushing them farther apart as the microtubule fibers lengthen. The sister chromatids begin to coil more tightly with the aid of condensin proteins and become visible under a light microscope.

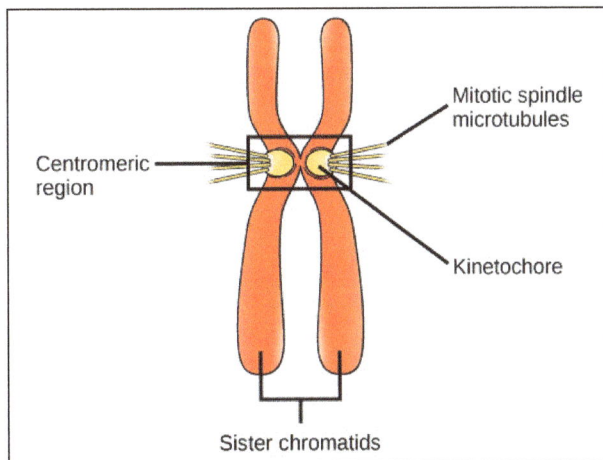

During prometaphase, mitotic spindle microtubules from opposite poles attach to each sister chromatid at the kinetochore. In anaphase, the connection between the sister chromatids breaks down, and the microtubules pull the chromosomes toward opposite poles.

During prometaphase, the "first change phase," many processes that were begun in prophase continue to advance. The remnants of the nuclear envelope fragment. The mitotic spindle continues to develop as more microtubules assemble and stretch across the length of the former nuclear area. Chromosomes become more condensed and discrete. Each sister chromatid develops a protein structure called a kinetochore in the centromeric region. The proteins of the kinetochore attract and bind mitotic spindle microtubules. As the spindle microtubules extend from the centrosomes, some of these microtubules come into contact with and firmly bind to the kinetochores. Once a mitotic fiber attaches to a chromosome, the chromosome will be oriented until the kinetochores of sister chromatids face the opposite poles. Eventually, all the sister chromatids will be attached via their kinetochores to microtubules from opposing poles. Spindle microtubules that do not engage the chromosomes are called polar microtubules. These microtubules overlap each other midway between the two poles and contribute to cell elongation. Astral microtubules are located near the poles, aid in spindle orientation, and are required for the regulation of mitosis.

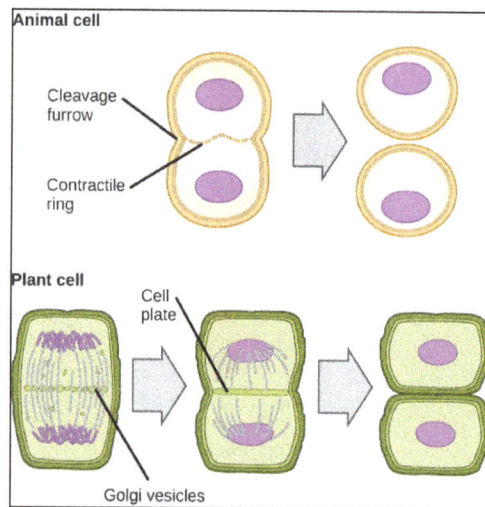

During metaphase, the "change phase," all the chromosomes are aligned in a plane called the metaphase plate, or the equatorial plane, midway between the two poles of the cell. The sister chromatids are still tightly attached to each other by cohesin proteins. At this time, the chromosomes are maximally condensed.

During anaphase, the "upward phase," the cohesin proteins degrade, and the sister chromatids separate at the centromere. Each chromatid, now called a chromosome, is pulled rapidly toward the centrosome to which its microtubule is attached. The cell becomes visibly elongated (oval shaped) as the polar microtubules slide against each other at the metaphase plate where they overlap.

During telophase, the "distance phase," the chromosomes reach the opposite poles and begin to decondense (unravel), relaxing into a chromatin configuration. The mitotic spindles are depolymerized into tubulin monomers that will be used to assemble cytoskeletal components for each daughter cell. Nuclear envelopes form around the chromosomes, and nucleosomes appear within the nuclear area.

Cytokinesis

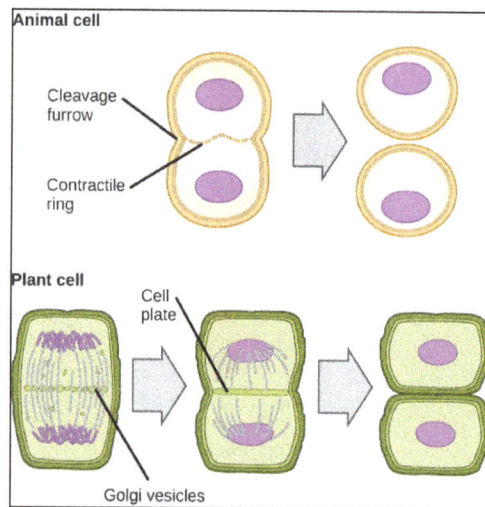

In figure, during cytokinesis in animal cells, a ring of actin filaments forms at the metaphase plate. The ring contracts, forming a cleavage furrow, which divides the cell in two. In plant cells, Golgi vesicles coalesce at the former metaphase plate, forming a phragmoplast. A cell plate formed by the fusion of the vesicles of the phragmoplast grows from the center toward the cell walls, and the membranes of the vesicles fuse to form a plasma membrane that divides the cell in two.

Cytokinesis, or "cell motion," is the second main stage of the mitotic phase during which cell division is completed via the physical separation of the cytoplasmic components into two daughter cells. Division is not complete until the cell components have been apportioned and completely separated into the two daughter cells. Although the stages of mitosis are similar for most eukaryotes, the process of cytokinesis is quite different for eukaryotes that have cell walls, such as plant cells.

In cells such as animal cells that lack cell walls, cytokinesis follows the onset of anaphase. A contractile ring composed of actin filaments forms just inside the plasma membrane at the former metaphase plate. The actin filaments pull the equator of the cell inward, forming a fissure. This fissure, or "crack," is called the cleavage furrow. The furrow deepens as the actin ring contracts, and eventually the membrane is cleaved in two.

In plant cells, a new cell wall must form between the daughter cells. During interphase, the Golgi apparatus accumulates enzymes, structural proteins, and glucose molecules prior to breaking into vesicles and dispersing throughout the dividing cell. During telophase, these Golgi vesicles are transported on microtubules to form a phragmoplast (a vesicular structure) at the metaphase plate. There, the vesicles fuse and coalesce from the center toward the cell walls; this structure is called a cell plate. As more vesicles fuse, the cell plate enlarges until it merges with the cell walls at the periphery of the cell. Enzymes use the glucose that has accumulated between the membrane layers to build a new cell wall. The Golgi membranes become parts of the plasma membrane on either side of the new cell wall.

G_0 Phase

Not all cells adhere to the classic cell cycle pattern in which a newly formed daughter cell immediately enters the preparatory phases of interphase, closely followed by the mitotic phase. Cells in G_0 phase are not actively preparing to divide. The cell is in a quiescent (inactive) stage that occurs when cells exit the cell cycle. Some cells enter G_0 temporarily until an external signal triggers the onset of G_1. Other cells that never or rarely divide, such as mature cardiac muscle and nerve cells, remain in G_0 permanently.

Control of the Cell Cycle

The length of the cell cycle is highly variable, even within the cells of a single organism. In humans, the frequency of cell turnover ranges from a few hours in early embryonic development, to an average of two to five days for epithelial cells, and to an entire human lifetime spent in G_0 by specialized cells, such as cortical neurons or cardiac muscle cells. There is also variation in the time that a cell spends in each phase of the cell cycle. When fast-dividing mammalian cells are grown in culture (outside the body under optimal growing conditions), the length of the cycle is about 24 hours. In rapidly dividing human cells with a 24-hour cell cycle, the G_1 phase lasts approximately nine hours, the S phase lasts 10 hours, the G_2 phase lasts about four and one-half hours, and the M phase lasts approximately one-half hour. In early embryos of fruit flies, the cell cycle is completed in about eight minutes. The timing of events in the cell cycle is controlled by mechanisms that are both internal and external to the cell.

Regulation of the Cell Cycle by External Events

Both the initiation and inhibition of cell division are triggered by events external to the cell when it is about to begin the replication process. An event may be as simple as the death of a nearby cell or as sweeping as the release of growth-promoting hormones, such as human growth hormone (HGH). A lack of HGH can inhibit cell division, resulting in dwarfism, whereas too much HGH can result in gigantism. Crowding of cells can also inhibit cell division. Another factor that can initiate cell division is the size of the cell; as a cell grows, it becomes inefficient due to its decreasing surface-to-volume ratio. The solution to this problem is to divide.

Whatever the source of the message, the cell receives the signal, and a series of events within the cell allows it to proceed into interphase. Moving forward from this initiation point, every parameter required during each cell cycle phase must be met or the cycle cannot progress.

Regulation at Internal Checkpoints

It is essential that the daughter cells produced be exact duplicates of the parent cell. Mistakes in the duplication or distribution of the chromosomes lead to mutations that may be passed forward to every new cell produced from an abnormal cell. To prevent a compromised cell from continuing to divide, there are internal control mechanisms that operate at three main cell cycle checkpoints. A checkpoint is one of several points in the eukaryotic cell cycle at which the progression of a cell to the next stage in the cycle can be halted until conditions are favorable. These checkpoints occur near the end of G_1, at the G_2/M transition, and during metaphase.

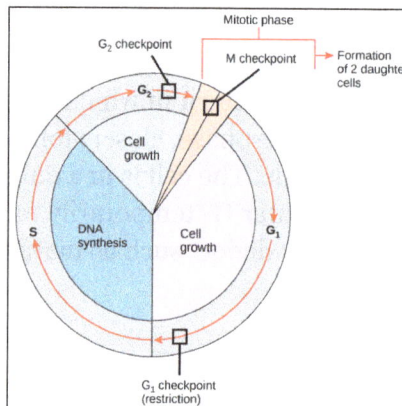

The cell cycle is controlled at three checkpoints. The integrity of the DNA is assessed at the G_1 checkpoint. Proper chromosome duplication is assessed at the G_2 checkpoint. Attachment of each kinetochore to a spindle fiber is assessed at the M checkpoint.

G_1 Checkpoint

The G_1 checkpoint determines whether all conditions are favorable for cell division to proceed. The G_1 checkpoint, also called the restriction point (in yeast), is a point at which the cell irreversibly commits to the cell division process. External influences, such as growth factors, play a large role in carrying the cell past the G_1 checkpoint. In addition to adequate reserves and cell size, there is a check for genomic DNA damage at the G_1 checkpoint. A cell that does not meet all the requirements will not be allowed to progress into the S phase. The cell can halt the cycle and attempt to remedy the problematic condition, or the cell can advance into G_0 and await further signals when conditions improve.

G_2 Checkpoint

The G_2 checkpoint bars entry into the mitotic phase if certain conditions are not met. As at the G_1 checkpoint, cell size and protein reserves are assessed. However, the most important role of the G_2 checkpoint is to ensure that all of the chromosomes have been replicated and that the replicated DNA is not damaged. If the checkpoint mechanisms detect problems with the DNA, the cell cycle is halted, and the cell attempts to either complete DNA replication or repair the damaged DNA.

M Checkpoint

The M checkpoint occurs near the end of the metaphase stage of karyokinesis. The M checkpoint is also known as the spindle checkpoint, because it determines whether all the sister chromatids are correctly attached to the spindle microtubules. Because the separation of the sister chromatids during anaphase is an irreversible step, the cycle will not proceed until the kinetochores of each pair of sister chromatids are firmly anchored to at least two spindle fibers arising from opposite poles of the cell.

Regulator Molecules of the Cell Cycle

In addition to the internally controlled checkpoints, there are two groups of intracellular molecules that regulate the cell cycle. These regulatory molecules either promote progress of the cell to the next phase (positive regulation) or halt the cycle (negative regulation). Regulator molecules may act individually, or they can influence the activity or production of other regulatory proteins. Therefore, the failure of a single regulator may have almost no effect on the cell cycle, especially if more than one mechanism controls the same event. Conversely, the effect of a deficient or non-functioning regulator can be wide-ranging and possibly fatal to the cell if multiple processes are affected.

Positive Regulation of the Cell Cycle

Two groups of proteins, called cyclins and cyclin-dependent kinases (Cdks), are responsible for the progress of the cell through the various checkpoints. The levels of the four cyclin proteins fluctuate throughout the cell cycle in a predictable pattern. Increases in the concentration of cyclin proteins are triggered by both external and internal signals. After the cell moves to the next stage of the cell cycle, the cyclins that were active in the previous stage are degraded.

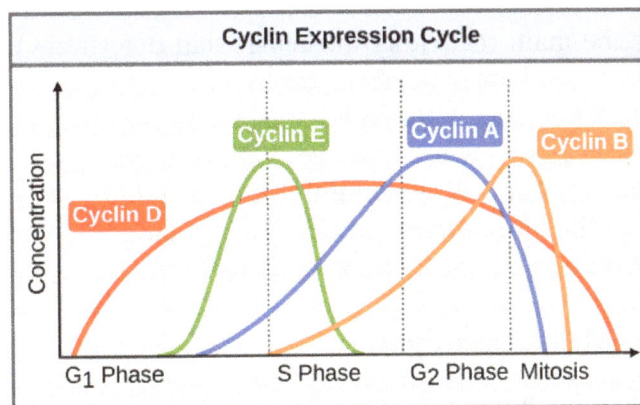

The concentrations of cyclin proteins change throughout the cell cycle. There is a direct correlation between cyclin accumulation and the three major cell cycle checkpoints. Also note the sharp decline of cyclin levels following each checkpoint (the transition between phases of the cell cycle), as cyclin is degraded by cytoplasmic enzymes.

Cyclins regulate the cell cycle only when they are tightly bound to Cdks. To be fully active, the Cdk/cyclin complex must also be phosphorylated in specific locations. Like all kinases, Cdks are enzymes (kinases) that phosphorylate other proteins. Phosphorylation activates the protein by changing its shape. The proteins phosphorylated by Cdks are involved in advancing the cell to the next phase. The levels of Cdk proteins are relatively stable throughout the cell cycle; however, the concentrations of cyclin fluctuate and determine when Cdk/cyclin complexes form. The different cyclins and Cdks bind at specific points in the cell cycle and thus regulate different checkpoints.

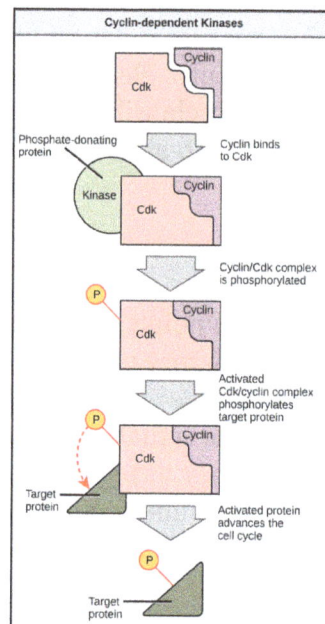

Cyclin-dependent kinases (Cdks) are protein kinases that, when fully activated, can phosphorylate and thus activate other proteins that advance the cell cycle past a checkpoint. To become fully activated, a Cdk must bind to a cyclin protein and then be phosphorylated by another kinase.

Since the cyclic fluctuations of cyclin levels are based on the timing of the cell cycle and not on specific events, regulation of the cell cycle usually occurs by either the Cdk molecules alone or the Cdk/cyclin complexes. Without a specific concentration of fully activated cyclin/Cdk complexes, the cell cycle cannot proceed through the checkpoints.

Although the cyclins are the main regulatory molecules that determine the forward momentum of the cell cycle, there are several other mechanisms that fine-tune the progress of the cycle with negative, rather than positive, effects. These mechanisms essentially block the progression of the cell cycle until problematic conditions are resolved. Molecules that prevent the full activation of Cdks are called Cdk inhibitors. Many of these inhibitor molecules directly or indirectly monitor a particular cell cycle event. The block placed on Cdks by inhibitor molecules will not be removed until the specific event that the inhibitor monitors is completed.

Negative Regulation of the Cell Cycle

The second group of cell cycle regulatory molecules are negative regulators. Negative regulators halt the cell cycle. Remember that in positive regulation, active molecules cause the cycle to progress.

The best understood negative regulatory molecules are retinoblastoma protein (Rb), p53, and p21. Retinoblastoma proteins are a group of tumor-suppressor proteins common in many cells. The 53 and 21 designations refer to the functional molecular masses of the proteins (p) in kilodaltons. Much of what is known about cell cycle regulation comes from research conducted with cells that have lost regulatory control. All three of these regulatory proteins were discovered to be damaged or non-functional in cells that had begun to replicate uncontrollably (became cancerous). In each case, the main cause of the unchecked progress through the cell cycle was a faulty copy of the regulatory protein.

Rb, p53, and p21 act primarily at the G_1 checkpoint. p53 is a multi-functional protein that has a major impact on the commitment of a cell to division because it acts when there is damaged DNA in cells that are undergoing the preparatory processes during G_1. If damaged DNA is detected, p53 halts the cell cycle and recruits enzymes to repair the DNA. If the DNA cannot be repaired, p53 can trigger apoptosis, or cell suicide, to prevent the duplication of damaged chromosomes. As p53 levels rise, the production of p21 is triggered. p21 enforces the halt in the cycle dictated by p53 by binding to and inhibiting the activity of the Cdk/cyclin complexes. As a cell is exposed to more stress, higher levels of p53 and p21 accumulate, making it less likely that the cell will move into the S phase.

Rb exerts its regulatory influence on other positive regulator proteins. Chiefly, Rb monitors cell size. In the active, dephosphorylated state, Rb binds to proteins called transcription factors, most commonly, E2F. Transcription factors "turn on" specific genes, allowing the production of proteins encoded by that gene. When Rb is bound to E2F, production of proteins necessary for the G_1/S transition is blocked. As the cell increases in size, Rb is slowly phosphorylated until it becomes inactivated. Rb releases E2F, which can now turn on the gene that produces the transition protein, and this particular block is removed. For the cell to move past each of the checkpoints, all positive regulators must be "turned on," and all negative regulators must be "turned off."

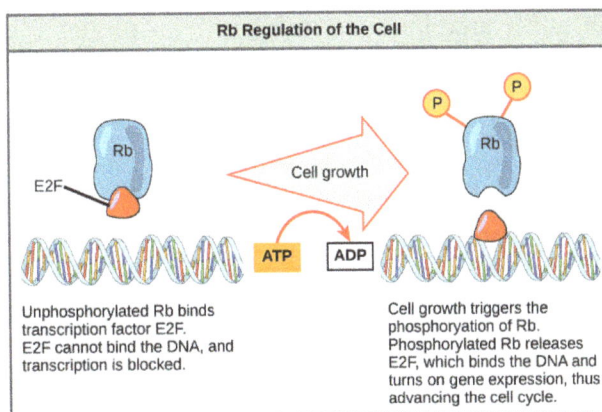

Rb halts the cell cycle and releases its hold in response to cell growth.

Rb and other proteins that negatively regulate the cell cycle are sometimes called tumor suppressors.

Cell Transport and Homeostasis

Probably the most important feature of a cell's phospholipid membranes is that they are selectively permeable. A membrane that is selectively permeable has control over what molecules or ions can enter or leave the cell, as shown in figure. The permeability of a membrane is dependent on the organization and characteristics of the membrane lipids and proteins. In this way, cell membranes help maintain a state of homeostasis within cells (and tissues, organs, and organ systems) so that an organism can stay alive and healthy.

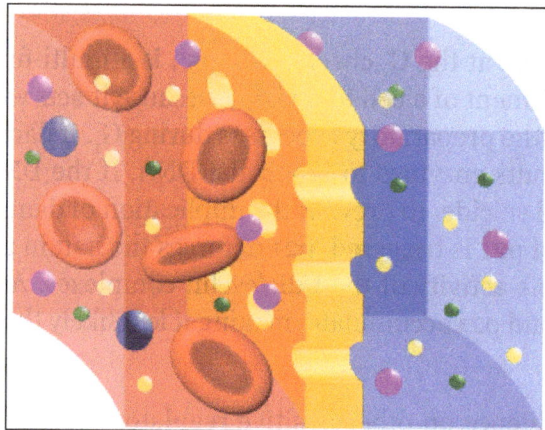

A selectively permeable membrane allows certain molecules through, but not others.

Transport across Membranes

The molecular make-up of the phospholipid bilayer limits the types of molecules that can pass through it. For example, hydrophobic (water-hating) molecules, such as carbon dioxide (CO_2) and oxygen (O_2), can easily pass through the lipid bilayer, but ions such as calcium (Ca^{2+}) and polar molecules such as water (H_2O) cannot. The hydrophobic interior of the phospholipid does not allow ions or polar molecules through because they are hydrophilic, or water loving. In addition, large molecules such as sugars and proteins are too big to pass through the bilayer. Transport proteins within the membrane allow these molecules to cross the membrane into or out of the cell. This way, polar molecules avoid contact with the nonpolar interior of the membrane, and large molecules are moved through large pores.

Every cell is contained within a membrane punctuated with transport proteins that act as channels or pumps to let in or force out certain molecules. The purpose of the transport proteins is to protect the cell's internal environment and to keep its balance of salts, nutrients, and proteins within a range that keeps the cell and the organism alive.

There are three main ways that molecules can pass through a phospholipid membrane. The first way requires no energy input by the cell and is called passive transport. The second way requires that the cell uses energy to pull in or pump out certain molecules and ions and is called active transport. The third way is through vesicle transport, in which large molecules are moved across the membrane in bubble-like sacks that are made from pieces of the membrane.

Passive Transport

Passive transport is a way that small molecules or ions move across the cell membrane without input of energy by the cell. The three main kinds of passive transport are diffusion, osmosis, and facilitated diffusion.

Diffusion is the movement of molecules from an area of high concentration of the molecules to an area with a lower concentration. The difference in the concentrations of the molecules in the two areas is called the concentration gradient. Diffusion will continue until this gradient has been eliminated. Since diffusion moves materials from an area of higher concentration to the lower, it is described as moving solutes" down the concentration gradient." The end result of diffusion is an equal concentration, or equilibrium, of molecules on both sides of the membrane.

If a molecule can pass freely through a cell membrane, it will cross the membrane by diffusion.

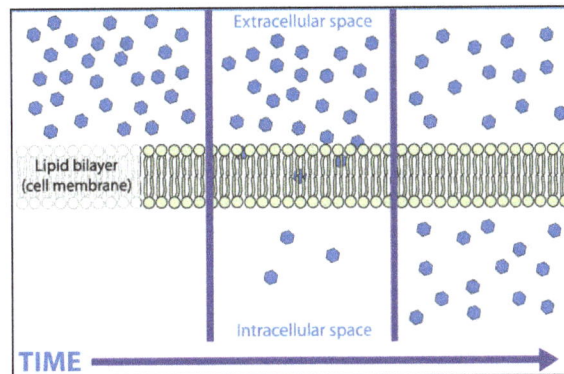

Molecules move from an area of high concentration to an area of lower concentration until an equilibrium is met. The molecules continue to cross the membrane at equilibrium, but at equal rates in both directions.

Osmosis

Imagine you have a cup that has 100ml water, and you add 15g of table sugar to the water. The sugar dissolves and the mixture that is now in the cup is made up of a solute (the sugar), that is dissolved in the solvent (the water). The mixture of a solute in a solvent is called a solution.

Imagine now that you have a second cup with 100ml of water, and you add 45 grams of table sugar to the water. Just like the first cup, the sugar is the solute, and the water is the solvent. But now you have two mixtures of different solute concentrations. In comparing two solutions of unequal solute concentration, the solution with the higher solute concentration is hypertonic, and the solution with the lower concentration is hypotonic. Solutions of equal solute concentration are isotonic. The first sugar solution is hypotonic to the second solution. The second sugar solution is hypertonic to the first.

We now add the two solutions to a beaker that has been divided by a selectively permeable membrane. The pores in the membrane are too small for the sugar molecules to pass through, but are big enough for the water molecules to pass through. The hypertonic solution is on one side of the membrane and the hypotonic solution on the other. The hypertonic solution has a lower water concentration than the hypotonic solution, so a concentration gradient of water now exists across the membrane. Water molecules will move from the side of higher water concentration to the side of lower concentration until both solutions are isotonic.

Osmosis is the diffusion of water molecules across a selectively permeable membrane from an area of higher concentration to an area of lower concentration. Water moves into and out of cells by osmosis. If a cell is in a hypertonic solution, the solution has a lower water concentration than the cell cytosol does, and water moves out of the cell until both solutions are isotonic. Cells placed in a hypotonic solution will take in water across their membrane until both the external solution and the cytosol are isotonic.

A cell that does not have a rigid cell wall (such as a red blood cell), will swell and lyse (burst) when placed in a hypotonic solution. Cells with a cell wall will swell when placed in a hypotonic solution, but once the cell is turgid (firm), the tough cell wall prevents any more water from entering the cell. When placed in a hypertonic solution, a cell without a cell wall will lose water to the environment, shrivel, and probably die. In a hypertonic solution, a cell with a cell wall will lose water too. The plasma membrane pulls away from the cell wall as it shrivels. The cell becomes plasmolyzed. Animal cells tend to do best in an isotonic environment, plant cells tend to do best in a hypotonic environment. This is demonstrated in figure.

When water moves into a cell by osmosis, osmotic pressure may build up inside the cell. If a cell has a cell wall, the wall helps maintain the cell's water balance. Osmotic pressure is the main cause of support in many plants. When a plant cell is in a hypotonic environment, the osmotic entry of water raises the turgor pressure exerted against the cell wall until the pressure prevents more water from coming into the cell. At this point the plant cell is turgid.

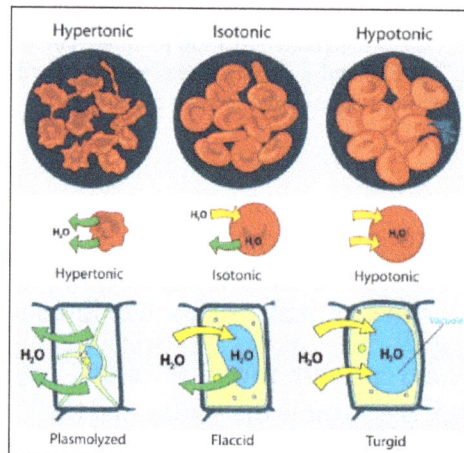

In figure, unless an animal cell (such as the red blood cell in the top panel) has an adaptation that allows it to alter the osmotic uptake of water, it will lose too much water and shrivel up in a hypertonic environment. If placed in a hypotonic solution, water molecules will enter the cell causing it to swell and burst. Plant cells (bottom panel) become plasmolyzed in a hypertonic solution, but tend to do best in a hypotonic environment. Water is stored in the central vacuole of the plant cell.

The effects of osmotic pressures on plant cells are shown in figure.

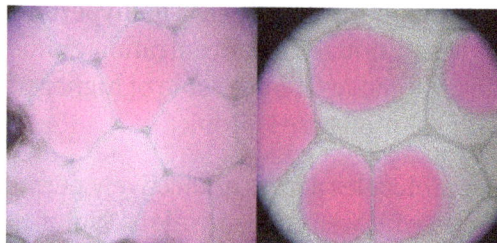

In figure, the central vacuoles of the plant cells in the left image are full of water, so the cells are turgid. The plant cells in the right image have been exposed to a hypertonic solution; water has left the central vacuole and the cells have become plasmolysed.

Osmosis can be seen very effectively when potato slices are added to a high concentration of salt solution (hypertonic). The water from inside the potato moves out of the potato cells to the salt solution, which causes the potato cells to lose turgor pressure. The more concentrated the salt solution, the greater the difference in the size and weight of the potato slice after plasmolysis.

The action of osmosis can be very harmful to organisms, especially ones without cell walls. For example, if a saltwater fish (whose cells are isotonic with seawater), is placed in fresh water, its cells will take on excess water, lyse, and the fish will die. Another example of a harmful osmotic effect is the use of table salt to kill slugs and snails.

Controlling Osmosis

Organisms that live in a hypotonic environment such as freshwater, need a way to prevent their cells from taking in too much water by osmosis. A contractile vacuole is a type of vacuole that removes excess water from a cell. Freshwater protists, such as the paramecia shown in figure, have a contractile vacuole. The vacuole is surrounded by several canals, which absorb water by osmosis from the cytoplasm. After the canals fill with water, the water is pumped into the vacuole. When the vacuole is full, it pushes the water out of the cell through a pore. Other protists, such as members of the genus Amoeba, have contractile vacuoles that move to the surface of the cell when full and release the water into the environment.

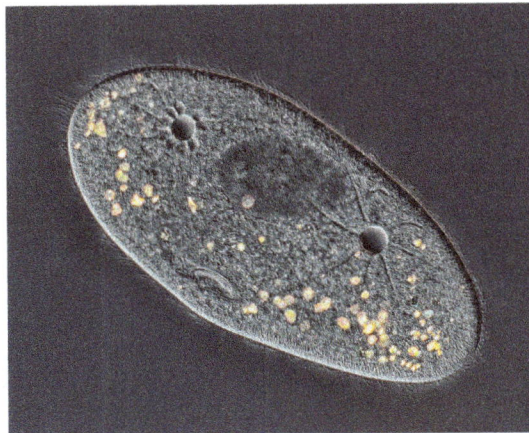

The contractile vacuole is the star-like structure within the paramecia (at center-right)

Facilitated Diffusion

Facilitated diffusion is the diffusion of solutes through transport proteins in the plasma membrane. Facilitated diffusion is a type of passive transport. Even though facilitated diffusion involves transport proteins, it is still passive transport because the solute is moving down the concentration gradient.

Small nonpolar molecules can easily diffuse across the cell membrane. However, due to the hydrophobic nature of the lipids that make up cell membranes, polar molecules (such as water) and ions cannot do so. Instead, they diffuse across the membrane through transport proteins. A transport protein completely spans the membrane, and allows certain molecules or ions to diffuse across the membrane. Channel proteins, gated channel proteins, and carrier proteins are three types of transport proteins that are involved in facilitated diffusion.

A channel protein, a type of transport protein, acts like a pore in the membrane that lets water molecules or small ions through quickly. Water channel proteins allow water to diffuse across the membrane at a very fast rate. Ion channel proteins allow ions to diffuse across the membrane.

A gated channel protein is a transport protein that opens a "gate," allowing a molecule to pass through the membrane. Gated channels have a binding site that is specific for a given molecule or ion. A stimulus causes the "gate" to open or shut. The stimulus may be chemical or electrical signals, temperature, or mechanical force, depending on the type of gated channel. For example, the sodium gated channels of a nerve cell are stimulated by a chemical signal which causes them to open and allow sodium ions into the cell. Glucose molecules are too big to diffuse through the plasma membrane easily, so they are moved across the membrane through gated channels. In this way glucose diffuses very quickly across a cell membrane, which is important because many cells depend on glucose for energy.

A carrier protein is a transport protein that is specific for an ion, molecule, or group of substances. Carrier proteins "carry" the ion or molecule across the membrane by changing shape after the binding of the ion or molecule. Carrier proteins are involved in passive and active transport. A model of a channel protein and carrier proteins is shown in figure.

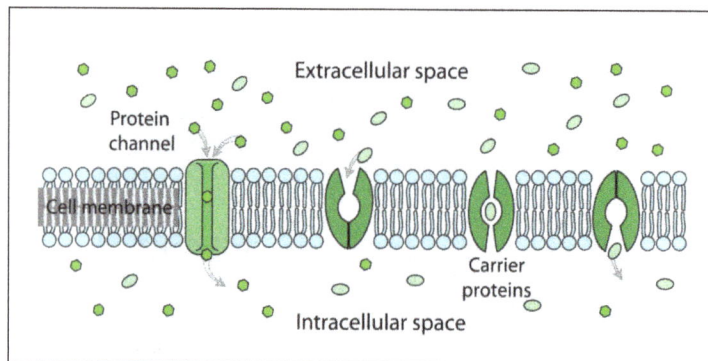

In figure, facilitated diffusion in cell membrane. Channel proteins and carrier proteins are shown (but not a gated-channel protein). Water molecules and ions move through channel proteins. Other ions or molecules are also carried across the cell membrane by carrier proteins. The ion or molecule binds to the active site of a carrier protein. The carrier protein changes shape, and releases the ion or molecule on the other side of the membrane. The carrier protein then returns to its original shape.

Ion Channels

Ions such as sodium (Na^+), potassium (K^-), calcium (Ca^{2+}), and chloride (Cl^-), are important for many cell functions. Because they are polar, these ions do not diffuse through the membrane. Instead they move through ion channel proteins where they are protected from the hydrophobic interior of the membrane. Ion channels allow the formation of a concentration gradient between the extracellular fluid and the cytosol. Ion channels are very specific as they allow only certain ions through the cell membrane. Some ion channels are always open, others are "gated" and can be opened or closed. Gated ion channels can open or close in response to different types of stimuli such as electrical or chemical signals.

Active Transport

In contrast to facilitated diffusion which does not require energy and carries molecules or ions down a concentration gradient, active transport pumps molecules and ions against a concentration gradient. Sometimes an organism needs to transport something against a concentration gradient. The only way this can be done is through active transport which uses energy that is produced by respiration (ATP). In active transport, the particles move across a cell membrane from a lower concentration to a higher concentration. Active transport is the energy-requiring process of pumping molecules and ions across membranes "uphill" against a gradient.

- The active transport of small molecules or ions across a cell membrane is generally carried out by transport proteins that are found in the membrane.

- Larger molecules such as starch can also be actively transported across the cell membrane by processes called endocytosis and exocytosis.

Sodium-potassium Pump

Carrier proteins can work with a concentration gradient (passive transport), but some carrier proteins can move solutes against the concentration gradient (from high concentration to low), with energy input from ATP. As in other types of cellular activities, ATP supplies the energy for most active transport. One way ATP powers active transport is by transferring a phosphate group directly to a carrier protein. This may cause the carrier protein to change its shape, which moves the molecule or ion to the other side of the membrane. An example of this type of active transport system, as shown in figure, is the sodium-potassium pump, which exchanges sodium ions for potassium ions across the plasma membrane of animal cells.

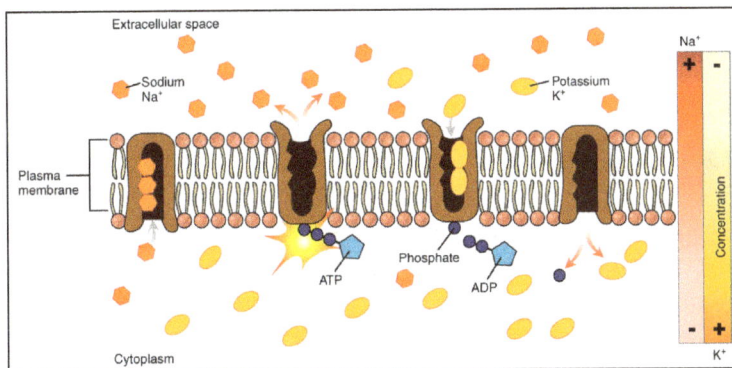

In figure, the sodium-potassium pump system moves sodium and potassium ions against large concentration gradients. It moves two potassium ions into the cell where potassium levels are high, and pumps three sodium ions out of the cell and into the extracellular fluid.

As is shown in figure, three sodium ions bind with the protein pump inside the cell. The carrier protein then gets energy from ATP and changes shape. In doing so, it pumps the three sodium ions out of the cell. At that point, two potassium ions move in from outside the cell and bind to the protein pump. The sodium-potassium pump is found in the plasma membrane of almost every human cell and is common to all cellular life. It helps maintain cell potential and regulates cellular volume. Cystic fibrosis is a genetic disorder that results in a misshapen chloride ion pump. Chloride levels within the cells are not controlled properly, and the cells produce thick mucus. The chloride ion pump is important for creating sweat, digestive juices, and mucus.

The Electrochemical Gradient

The active transport of ions across the membrane causes an electrical gradient to build up across the plasma membrane. The number of positively charged ions outside the cell is greater than the number of positively charged ions in the cytosol. This results in a relatively negative charge on the inside of the membrane, and a positive charge on the outside. This difference in charges causes a voltage across the membrane. Voltage is electrical potential energy that is caused by a separation of opposite charges, in this case across the membrane. The voltage across a membrane is called membrane potential. Membrane potential is very important for the conduction of electrical impulses along nerve cells.

Because the inside of the cell is negative compared to outside the cell, the membrane potential favors the movement of positively charged ions (cations) into the cell, and the movement of negative ions (anions) out of the cell. So, there are two forces that drive the diffusion of ions across the plasma membrane—a chemical force (the ions' concentration gradient), and an electrical force (the effect of the membrane potential on the ions' movement). These two forces working together are called an electrochemical gradient.

Vesicles and Active Transport

Some molecules or particles are just too large to pass through the plasma membrane or to move through a transport protein. So cells use two other methods to move these macromolecules (large molecules) into or out of the cell. Vesicles or other bodies in the cytoplasm move macromolecules or large particles across the plasma membrane. There are two types of vesicle transport, endocytosis and exocytosis.

Endocytosis and Exocytosis

Endocytosis is the process of capturing a substance or particle from outside the cell by engulfing it with the cell membrane. The membrane folds over the substance and it becomes completely enclosed by the membrane. At this point a membrane-bound sac, or vesicle pinches off and moves the substance into the cytosol. There are two main kinds of endocytosis:

- Phagocytosis or "cellular eating," occurs when the dissolved materials enter the cell. The plasma membrane engulfs the solid material, forming a phagocytic vesicle.

- Pinocytosis or "cellular drinking," occurs when the plasma membrane folds inward to form a channel allowing dissolved substances to enter the cell, as shown in figure. When the channel is closed, the liquid is encircled within a pinocytic vesicle.

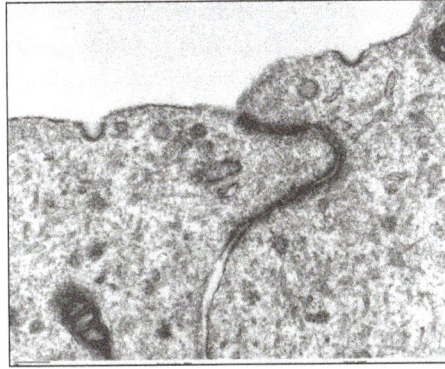

Transmission electron microscope image of brain tissue that shows pinocytotic vesicles. Pinocytosis is a type of endocytosis.

Exocytosis describes the process of vesicles fusing with the plasma membrane and releasing their contents to the outside of the cell, as shown in figure. Exocytosis occurs when a cell produces substances for export, such as a protein, or when the cell is getting rid of a waste product or a toxin. Newly made membrane proteins and membrane lipids are moved on top the plasma membrane by exocytosis.

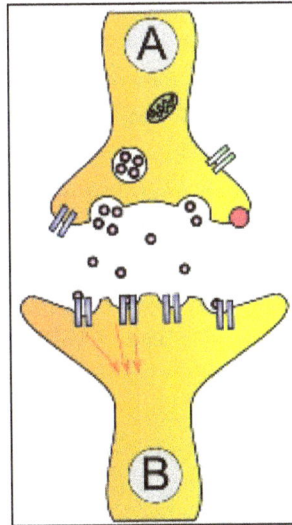

Mode of exocytosis at a synaptic junction, where two nerve cells meet. Chemical signal molecules are released from nerve cell A by exocytosis, and move toward receptors in nerve cell B. Exocytosis is an important part in cell signaling.

Homeostasis and Cell Function

Homeostasis refers to the balance, or equilibrium within the cell or a body. It is an organism's ability to keep a constant internal environment. Keeping a stable internal environment requires constant adjustments as conditions change inside and outside the cell. The adjusting of systems within a cell is called homeostatic regulation. Because the internal and external environments of a cell are constantly changing, adjustments must be made continuously to stay at or near the set point (the normal level or range). Homeostasis is a dynamic equilibrium rather than an unchanging state. The cellular processes discussed all play an important role in homeostatic regulation.

Cell Migration

Cell migration is fundamental to establishing and maintaining the proper organization of multicellular organisms. Morphogenesis can be viewed as a consequence, in part, of cell locomotion, from large-scale migrations of epithelial sheets during gastrulation, to the movement of individual cells during development of the nervous system. In an adult organism, cell migration is essential for proper immune response, wound repair, and tissue homeostasis, while aberrant cell migration is found in various pathologies. Indeed, as our knowledge of migration increases, we can look forward to, for example, abating the spread of highly malignant cancer cells, retarding the invasion of white cells in the inflammatory process, or enhancing the healing of wounds.

Types of Cell Migration and Related Phenomena

Fibroblasts

In vivo, fibroblasts are typically found in connective tissue where they synthesize collagens, glycosaminoglycans, and other important glycoproteins of the extracellular matrix (ECM) including fibronectin, for example. *In vitro*, these cells have been objects of extensive study because of the ease of culturing them. Fibroblasts cultured on glass have a spread or spindle-shaped morphology, often characterized by several extending processes. In cell culture, fibroblasts move slowly with an average speed less than 1 μm/min and often change direction. It is from fibroblast cell migration that the textbook paradigm for the classic steps of locomotion is derived. The locomotory cycle consists of cells protruding and subsequently adhering at the leading margin, developing contractile forces between the front and trailing margins, and finally releasing trailing adhesions due to the applied tension and enzymatic action. Retraction generates excess dorsal surface to sustain the protrusion in a process termed retraction induced spreading. Over the past several decades, considerable work has been devoted to understanding the mechanistic steps of cell migration as exemplified by fibroblasts.

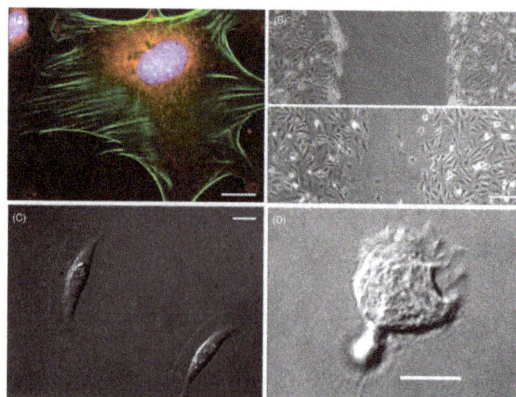

In figure, different types of cell migration. (A) A stationary, spread C3H10T1/2 fibroblast triple stained with DAPI (blue) for DNA, MitoTracker (red) for mitochondria, and Alexa Fluor phalloidin for F-actin. (B) Fibroblasts migrating into wound. Top: initially, a wound was made in a confluent monolayer of MDA-MB-231cells by scratching using a pipette tip. Bottom: after 15 h, migrating cells began to fill in the wound. (C) Migrating zebrafish keratocytes with large fan-like lamellipodia. (D) An HL-60 cell (human promyelocytic leukemia cell) migrating on a glass substrate after differentiation with dimethyl sulfoxide (DMSO) to exhibit leukocyte-like behavior on glass substrate. (Image in 1A and 1D are courtesy of Bing Yang and Zenon Rajfur, respectively.) Scale bars in A, C, and D are 10 µm, in B is 100 um.

Fibroblasts play a critical role in wound healing. *In vivo*, and *in vitro*, fibroblasts migrate into wounds, in the process cell acquiring cues that enable them to secrete ECM proteins and proliferate. However, they migrate *in vitro* with different speeds and morphology when compared to single fibroblasts in cell culture. Fibroblasts migrating into a wound tend to have a large lamellipodium extending into the wound with few stress fibers in the cell; by contrast, stationary fibroblasts have smaller lamellipodia, and are characterized by multiple stress fibers. A typical wound healing assay is shown in figure. It is known that many of the growth factors presented at a wound site act either as mitogens or as chemotactic factors for fibroblasts; these include, for example, epidermal growth factor (EGF) and platelet-derived growth factor (PDGF). Stimulation by growth factors can increase single fibroblast migration speed up to 3-fold, at the same time increasing changes in cell migration direction.

Keratocytes

At the other end of the spectrum of cell locomotion, fish or amphibian keratocytes migrate in rapid, highly persistent mode in which protrusion, contraction, and retraction are smoothly coordinated so that the cell maintains a nearly constant shape. Keratocytes are terminally differentiated epithelial cells in fish and amphibians that make good models for several aspects of migrating cells. In primary cultures of scales, keratocytes from goldfish were found to move away from the scale with high velocities (typically 10–15 µm/min but occasionally up to 60 µm/min). The highly directional movement of isolated keratocytes may originate from their ability to move as sheets to close wounds at the surface of the scale. Indeed, they are robust migration machines, migrating for days under proper culture conditions. Even keratocytes lacking the nucleus and microtubules (MTs) can migrate following a stimulus.

Lamellipodium Structure

Keratocytes have a large fan-like lamellipodium. The cell body at the base of lamellipodium is pulled (laterally) into to an elongated shape by actin bundles; in keratocytes MTs and intermediate filaments do not penetrate the thin, actin-rich lamellipodium but are confined to the perinuclear region. Light, fluorescence, and electron microscope images of f-actin in the lamellipodium can be reconciled and all show an oriented f-actin network; this presumably reflects the underlying branched actin network as described by the dendritic nucleation model. However, the issue of the predominant f-actin structure is not completely settled, and an alternate view is offered.

Cytoskeletal Dynamics and Migration

Considerable work has been devoted to the cytoskeletal mechanisms involved in keratocyte migration Actin polymerization, treadmilling, retrograde actin network flow, and myosin II-based contractility all play major roles in migrating keratocytes. Indeed, the force required to stall a protruding keratocyte is consistent with an actin polymerization ratchet model; however, the shape of the force-velocity curve is indicating additional factors come into play when the elastic ratchet model is placed in a cellular context. Careful examination of actin flows using fluorescence speckle microscopy (FSM) reveals retrograde actin flow, smaller at the leading edge and larger at the wings (sides) of the keratocyte. The difference between protrusion and retrograde actin flow rates represents the net actin polymerization rate that is highest at center of the leading margin and falls off toward the wings. These flows are related to tractions exerted on the substratum.

Interestingly, the myosin II network moves relative to the actin network. Since the myosin II inhibitor, blebistatin, reduces keratocyte locomotion, cell body translocation involves both actomyosin contraction as well as actin assembly. In fact, Theriot and co-workers demonstrated a novel role for myosin II in addition to its well-known role powering contraction: by accelerating network disassembly, myosin II activity leads to network shrinkage via tension induced actin filament breakage. This action will not only directly lead to retraction but it also recycles monomeric actin for new polymerization at the front.

One effect of myosin II-based contraction is to drive a forward flow of cytoplasm in migrating keratocytes. By measuring the front to rear gradient (higher in the front) in the concentration of quantum dots that had been introduced into the cytoplasm and fitting this data to a simple model for flow driven accumulation at the front, anterograde flow velocities in the cell frame of reference that were about 1/3 that of the keratocyte velocity (~0.1 µm/s vs. ~0.3 µm/s) were obtained. Such flows could augment migration by feeding more actin monomer to the growing network at the leading edge and perhaps even providing pressure on the cell surface at the leading margin making network growth via actin polymerization more facile.

Shape and Migration

Based on a shape and speed analysis of hundreds of cells, Theriot and co-workers proposed a model for observed keratocyte morphology and crawling behavior. Their model is based on the notion that actin polymerization and treadmilling drives migration but is it is resisted by the constant tension of an inextensible membrane surrounding cells of constant area. Spatial differences in the density of growing actin filament network, namely, that the density of filaments is graded with highest values at the center of the leading edge, give rise to characteristic shape of the dominant modes of keratocyte locomotion. Thus, cells with higher actin density at the center than at the sides will have a larger aspect ratio defined as the ratio of the long axis (width) to short axis (length) of the keratocyte. In this model, global integration of spatially varying actin polymerization powered protrusion is provided by membrane tension to specify cell shape. In addition, the model predicts that cell speed will be positively related to the aspect ratio of the cells; thus, canoe-shaped keratocytes, with a larger aspect ratio, move faster than D-shaped cells, with a smaller aspect ratio.

Mogilner and colleagues have constructed in silico models of keratocyte locomotion in which several qualitative notions are incorporated mathematically. At the front of the cell, the dendritic nucleation model is responsible for protrusion while at the rear, the dynamic network contraction model is responsible for retraction. Recently, a model of a viscoelastic lamellipodium was generated using a realistic geometry that correctly predicts measured centripetal flow of the actin network and the positive gradient of myosin II going from front to rear.

Leukocytes

Leukocytes, or white blood cells (WBCs), are cells of the immune system defending the body against infecting organisms and foreign materials. They are highly motile cells found throughout the body, including tissues, blood, and the lymphatic system. The recruitment of leukocytes to the site of bacterial and viral infection involves initial attachment to vascular endothelium, rolling, weak and firm adhesion, transendothelial migration, and chemotaxis. Leukocyte chemotaxis *in vivo* and *in vitro* occurs at speeds around 4 μm/min. Leukocytes migrate on different substrates through adhesions that involve the integrins β_2 and $\alpha_4\beta_1$. However, recently it has been reported that leukocytes can adhere and migrate in an integrin-independent manner, indicating that leukocytes employ additional mechanisms for adhesion and migration. A view of differentiated migrating HL-60 leukemia-like cell.

Single-cell Migration in Three Dimensions

Although cell migration has been studied extensively in essentially 2D cell culture conditions where cells grow on a substrate, increasing attention has been paid to the movement of cells in 3D environments. The 3D matrix acts as a scaffold that produces physical support for cells that can affect cell morphology and induce cell growth or migration. In addition, the matrix can induce variation in signaling cascades in cells via adhesions and tensile forces.

Cell Morphology and Migration in 3D Environments

Most migration modes previously observed in 2D environments also occur in 3D tissue environments. However, because the distribution of ligands in 2D is generally much more uniform than in 3D matrix models where, for example, clustered ligands may exist on fibrils, cell morphology is quite different in the two environments. In 2D cell culture, fibroblasts have large lamellipodia and filopodia. By contrast, fibroblasts in 3D collagen gels exhibit both smaller and fewer lamellipodia and filopodia. Due to extensive adhesion to a flat substratum, cells in 2D show very broad, flat and thin lamellipodia whereas cells in 3D show a less exaggerated appearance. Three motile morphologies can be delineated in a 3D matrix: amoeboid blebby (macrophages, some stem cells on soft/loose connective tissue); amoeboid pseudopodal (leukocytes, dictyostelium on loose connective tissue); and mesenchymal (fibroblasts, and some cancer cells on loose or dense connective tissue).

Regulation of Cell Migration in 3D Matrices

Three important factors regulate 3D cell migration: cell-matrix adhesions, the Rho family of small GTPases, and proteases. In 2D culture, integrins are primarily responsible for cell adhesions to ECM in the form of focal adhesions (FAs), focal contacts, podosomes, etc. However, in 3D cell culture, a reduction in the number of FAs and their component integrins occurs. Thus, for example, αVβ3 integrin, which is highly expressed in 2D cell culture, was not detected in the 3D-matrix adhesions of fibroblasts, and the level of FA kinase (FAK) phosphorylation was reduced. Changes in the nature and strength of adhesions in 3D and 2D environments will result in differences in cell tension, morphology, and migration type.

The Rho family of small GTPases plays a prominent role in regulating cell migration in 3D. Leukocytes employ amoeboid migration that is based on the Rho/Rho-associated protein kinase (ROCK) pathway maintaining contractility at the posterior end and Rac1 mediating protrusion at the leading margin. However, other reports indicate that Rac1 activity is suppressed in fibroblasts and neurons in 3D culture, thus decreasing leading edge ruffling and axonal branching, respectively.

The role of proteolysis in 3D migration in tissue has been actively investigated. Multiple proteases have collagenolytic activity but the emphasis has been on matrix-metallo proteases (MMPs) and these have been reported to affect both normal and cancer cell migration *in vitro*. However, clinical trials of MMP inhibitors did not impair metastasis suggesting that metastatic cells may switch from mesenchymal to amoeboid locomotion.

Adhesions in Migrating Cells

Cells adhere to ECM or other cells by both nonspecific electrostatic interactions and specific binding of cell adhesion molecules such as selectins, integrins, and cadherins to ECM ligands and to cadherins on other cells. We will focus on cell-ECM adhesions, and divide such adhesions into FAs, podosomes, focal complexes, and close contacts.

Focal Adhesions: Composition and Structure

FAs were first identified in chicken heart fibroblasts by electron microscopy, as dense plaques between the cell's ventral surface and the substrate. FAs are usually found at the ends of stress fibers; they have a dimension on the order of a micron, and a lifetime ranging between minutes and hours. They have been visualized by epifluorescence microscopy, by total internal fluorescence microscopy (TIRFM), or by interference reflection contrast microscopy (IRM). In the past, terms such as adhesion plaques, or focal contacts were employed, but now the field appears to have settled on the term FA. FA components can be divided into four general categories: (1) ECM components, of which fibronectin, laminin, vitronectin, and the collagens are important examples; (2) transmembrane proteins, of which integrins are the most prominent class; (3) structural proteins that both stabilize the FA and provide scaffolding functions; and (4) signaling proteins. The number of proteins found in FAs is now exceeds 160, and the possible interactions between these components is described in what is colloquially called the "Geiger diagram" which evolves as new components are identified.

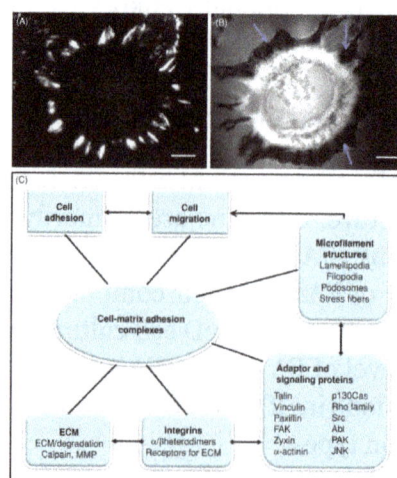

In figure, Adhesion structure and function in cells. (A) An immunofluorescence image of focal adhesions (FAs) in an NIH 3T3 cell stained with antipaxillin; (B) an interference reflection microscopy (IRM) image of FAs in a similar NIH 3T3 fibroblast on a fibronectin (FN)-coated substrate; the very dark regions (arrows) are FAs; and (C) schematic figure for the relationship between cell adhesion, cell migration, and some of the corresponding adaptor and signal proteins. Cell matrix adhesion complexes are depicted a key component in single-cell adhesion and migration. After activation, integrins bind extracellular matrix (ECM) and provide a link to the actin cytoskeleton. Cytoplasmic adaptor proteins bind integrin cytoplasmic domains, stabilize FA, and provide scaffolding functions. Integrin activation also initiates downstream signaling. Such signaling may regulate cell adhesion turnover, internal force development, and cytoskeletal rearrangements including formation of stress fibers, lamellipodia, filopodia, and podosomes. Cell migration also involves both ECM degradation and proteolysis and adhesion complex internalization. Scale bars in A and B are 10 μm.

Integrins are the transmembrane proteins that recognize ECM proteins containing short amino acid sequences, such as the arginine-glycine-aspartic acid (RGD), Asp-Gly-Glu-Ala (DGEA), and glycine-phenylalanine-hydroxyproline-glycine-glutamate-arginine (GFOGER) motifs. Functional integrins are heterodimers containing two distinct (α and β) subunits. Currently, there are more than 24 types of α and β integrin subunits characterized in mammals. Each type of integrin heterodimer binds distinct ligands, for example, α5β1 integrin binds fibronectin, and α3β1 bind to laminin. FAs in different fibroblasts and epithelial cells that are adherent to distinct ECM materials contain integrins with various combinations of α- and β-subunits.

One function of cytoskeletal proteins, including talin, α-actinin, filamin, and tensin, is to link integrins to the actin cytoskeleton. Other adaptor proteins directly or indirectly interact with integrin cytoplasmic tails and form protein complexes; examples include FAK, vin-culin (Vn), paxillin (Pax), dynamin, and Ena/vasodilator-stimulated phosphoprotein (VASP). As an example, an epifluorescence image of antibody labeled Pax is given in figure and shows the extensive array of FAs in murine fibroblasts adherent to a serum coated glass substrate.

Signaling proteins are recruited to FA and regulate their assembly and disassembly; examples include the Src family of nonreceptor tyrosine kinases (NRPTKs), the Abl family NRPTK and the Rho family of small GTPases, and p21-activated kinase. In addition, phospho-rylation of Pax by c-Jun amino-terminal kinase (JNK) or cdk 5 has been found essential for maintaining the labile adhesions required for rapid migration in both fibroblasts and neurons. Some proteins and signaling pathways involved in FA structure and regulation and their relationship to cell adhesion and migration are diagrammed schematically.

FA appears to be an amorphous collection of interacting proteins making 3D structure determinations difficult be either light or electron microscopy. However, recently progress has been made employing photoactivation localization microscopy (PALM) in 2D and by iPALM, in 3D. Such studies are revealing the 3D organization of individual FA proteins.

Focal Adhesion Dynamics

FAs are dynamic structures that undergo cycles of assembly and disassembly; indeed, regulated FA turnover is integral to cell migration.

Focal Adhesion Assembly

The role of integrin activation in FA assembly and in initiating downstream signaling has been extensively investigated. With stimulation, for example, by growth factors, integrin β-subunit cytoplasmic domains bind the talin phosphotyrosine-binding (PTB) domain causing integrin activation. Activated integrins then bind ECM components and the cytoplasmic domain recruits signaling proteins; this process initiates downstream signaling, including FAK phosphorylation, mitogen-activated protein kinase (MAPK) activation, Pax binding, and the formation of a complex containing Vn, FAK, α-actinin, Wiskott-Aldrich syndrome protein (WASP), tensin, Src, and zyxin. Knockouts of key recruited signaling components have demonstrable effects on cell adhesion and migration. Thus, for example, FAK null fibroblasts exhibit increased numbers of adhesions and consequent reduced cell motility. In addition, kinase dead Src mutants promoted both the number and size of cell adhesions, reducing the speed of cell migration. Webb et al. found that Src, Pax, and FAK formed complexes *in vitro* and *in vivo*; in this study, FAK and Src were speculated to regulate cell adhesion disassembly via Pax and the downstream extracellular-signal-regulated kinase (ERK) and myosin light-chain kinase (MLCK) pathways. Abl knockdown cells also exhibited an increase in cell adhesion size and stability, and rescue of Abl kinase activity restored the cell adhesion disassembly rate. Rho family GTPases have also been reported as key regulators of FA dynamics, for example, active RhoA changed small peripheral adhesions (focal complexes) into elongated FAs. External stretch induced nascent adhesions to mature into FAs via a RhoA-ROCK pathway.

Focal Adhesion Disassembly

Compared with extensive studies on FA formation, the disassembly process is not as clear. Several related pathways may contribute to FA disassembly: (i) adhesion release produced by ECM degradation; (ii) adhesion turnover mediated by the cytoskeleton and internalization; and (iii) disassembly mediated by kinases and proteases. It has been reported that ECM degradation is, in part, responsible for cell adhesion disassembly, cell migration, and invasion; thus, for example, ECM degradation by matrix metalloproteinases (MMPs) could induce the release of cell adhesions resulting in an increase cell motility and invasion.

Cytoskeletal components are an important regulatory factor in adhesion disassembly. MTs have been observed to target FAs promoting their disassembly. Moreover, MTs have been speculated to induce cell adhesion disassembly via dynamin- and clathrin-dependent integrin endocytosis. Caveolin-1 was also reported to regulate FA turnover and cell migration directionality possibly via internalization. In addition, cellular contractile machinery may also induce FA disassembly; for example, RhoA, and myosin II were found to positively regulate adhesion disassembly and cause cell rear detachment.

Proteases and kinases have also been reported to regulate cell adhesion. Calpain, a calcium-dependent protease cleaves talin, FAK, and Pax in FA. Cleavage of these proteins leads to disassembly of the FA and the detachment of the tail of the cell. Moreover, recent studies have demonstrated that, Smurf1, an E3 ubiquitin ligase, degrades the talin head and controls cell adhesion stability. Other ubiquitin ligases, including Cbl, Smurf2, HDM2, and BCA2, also play an important role in regulating cell adhesion and migration through ubiquitination of their specific substrates.

Methods have been developed to study the dynamics of FAs. Studies using fluorescence recovery after photobleaching (FRAP) and green fluorescent protein (GFP)-fusion proteins or labeled microinjected proteins have shown that protein components of FAs including α-actinin, Vn, and FAK slowly exchange between the cytosol and the adhesion with half-times for recovery on the order of minutes. More recently, Horwitz and co-workers measured adhesion disassembly rates of FP conjugated Pax, FAK, and zyxin; these studies indicated that the FAK-Src complex could interrupt FA maturation by promoting disassembly through the downstream ERK and MLCK pathways. Using the techniques of image correlation microscopy, Gratton, Wiseman, Horwitz, and their co-workers measured FAK, Vn, and Pax diffusion and binding to adhesions in mouse embryonic fibroblasts. No FAK, Vn, and Pax complexes were preassembled in cytoplasm, but when the adhesions disassembled, these proteins disassociated in complexes. Waterman and colleagues studied FA dynamics using speckle microscopy and advanced image analysis; they found that the retrograde F-actin network velocity is a fundamental regulator of traction force at FAs via the Rho and myosin II pathways. These investigators also demonstrated that the interplay between actomyosin and FA dynamics results in a balance between adhesion and contraction to induce maximal migration velocity. Such studies indicated a relationship between force and FA assembly and disassembly, and predicted how under certain circumstances the FA slide.

Podosomes

Podosomes are specialized integrin-mediated adhesions often found in highly migratory monocytic cells that mediate the inflammatory response. They also have the capacity for matrix degradation. Linking the ECM to the actin cytoskeleton, podosomes have a fairly uniform dimension of around 0.5 μm, a half-life of 2 to 20 min and are abundant (20–100 per cell). An image of podosomes is shown in figure.

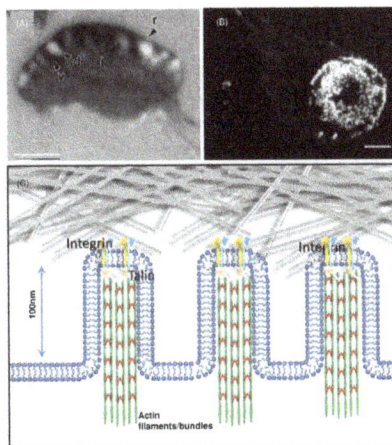

In Figure, (A) Interference reflection microscopy (IRM) image of close adhesion in migrating fish kera-tocytes, the adhesion pattern consists of an outer rim (r) of very close contact skirting a crescent-shaped band of alternating very close (v) and distant contacts (d). (B) Epifluorescent image of podosomes in a human dendritic cell with F-actin labeling. (C) A hypothetical view of close contacts in which small diameter projections attach to the substrate and serve to draw the ventral surface closer to the substrate such that it appears gray in IRM. Integrin, talin, F-actin have been reported to be in close adhesions [in this schematic, the actin network is depicted like that in a microvillus with parallel actin bundles but it could also be in the form of a dendritic actin network (not shown)]; however, paxillin and focal adhesion kinase (FAK) are not found in initial close contacts. Scale bars are 10 μm. Image in panel A is from Lee and Jacobson; image in panel B is courtesy of Aaron Neumann.

Podosomes have a dense actin core surrounded by a rosette-like structure containing integrins, such as $\alpha_v\beta_3$, FA proteins including talin and Vn that play a major structural role, other actin-associated proteins [gelsolin, alpha-actinin, and actin-related protein 2/3 (Arp2/3)], tyrosine kinases (Src, Pyk2), and phosphoinositide-3 kinase (PI3K), and also the Rho-family GTPases. The podosome core also contains proteins involved in regulating actin polymerization including WASP.

Focal Complexes

The term, "focal complex," describes small adhesions that form at the leading margin of migrating cells, typically fibroblasts. Focal complexes are significantly smaller in area (<0.25 μm²), and are shorter lived (often <5 min but some have even shorter lifetimes) than FAs. Focal complexes contain integrins, talin, and Pax, but fewer actin filaments are associated with them. Migrating cells often have a large number of focal complexes at the protruding edge. Most of these focal complexes never mature, and are likely disassembled when the lamellipodium retracts. Some investigators have suggested that focal complexes might be precursors of FA because applied contractile forces can convert focal complexes into larger oval-shape adhesions.

Close Contacts in Migrating Cells

Close contacts appear as broad gray areas in IRM. The original definition of close contacts was based on IRM images and indicated that the separation between the ventral surface of the cell and the substratum was about 20 to 50 nm. By contrast, the ventral surface and substratum is separated by 10 to 15 nm or less in FA. Compared to FA, little is known about these adhesions. They predominate in fast moving cells such as keratocytes although regions of close contact also exist in fibroblasts and epithelial cells in culture.

The composition of close contacts was investigated by immunofluorescence staining of fish keratocytes using antibodies against known FA components. The close contact areas at the rim of leading edge were found enriched in β1-integrin and talin, with little Pax and FAK. In general, close contacts appear to be mediated by integrins. Forward movement of the Xenopus keratocyte lamella could be halted by adding RGD peptide or an anti-integrin mAb while the rear of cell continued to retract.

Anderson and Cross performed a detailed study of more mature Vn-containing adhesions using microinjected fluorescent Vn and combined confocal and IRM imaging. They found that these contacts formed behind the leading edge and matured beneath the lamellipodium and remained stationary while the cell passed over them. By contrast, Vn-containing contacts in the wings of the cell grew larger before sliding inward. These large contacts are presumably transmitting the large lateral traction in keratocytes that are used for retraction of the wings. The actual mechanism for disassembly of released contacts remains an open question.

There are really no structural models for close contacts. A possible model would consist of finger-like projections of a small diameter that contact the surface using the usual repertoire of FAs molecules. In this respect, these projections would be a cross between podosomes and filopodia. The net result would be to draw the surface closer to the substratum such that the region appears gray in IRM yet the adhesion itself could be readily remodeled to accommodate rapid cell migration.

Outlook

In addition to the extensive cataloging of adhesion components, there are recent developments in super-resolution microscopy and several live cell fluorescence microscopy methods that promise to enhance our understanding of structure-function relationships in the adhesive structures that enable the cell to exert traction on its environment. Also, recent developments in Rho family biosensors and detailed analysis of such data, promise to provide detailed mapping of the localization and activation pattern of these GTPases in relation to the regulation of dynamic adhesive behavior, tractions, and cell migration. Overall, it appears that the next decade will produce important advances in our understanding of cell-substratum adhesions.

Measurements of Tractions in Single Migrating Cells

Elastic Substrate Traction Measurements

The effects of tractions exerted by migrating chick heart fibroblasts plated on a deformable silicone substrate (a thin film of silicone cross linked by means of glow discharge) were visualized as visible wrinkles in the film under the cell body and perpendicular to the direction of cell movement. Such compression wrinkles qualitatively reflect the strong contractile forces exerted by fibroblasts on their environment but do not give the actual distribution of traction stresses under the cell.

Spatially resolved information on the distribution of tractions has been obtained in the past 15 years by following the displacements of fiduciary markers embedded in deformable substrata or the response of individual force-sensing elements. This approach was first applied to fish scale keratocytes migrating on silicone rubber substrata in which small polystyrene latex beads had been embedded. When the tractions were calculated from the bead displacements, it was found, surprisingly, that the major propulsive tractions were applied in the wings of the keratocyte.

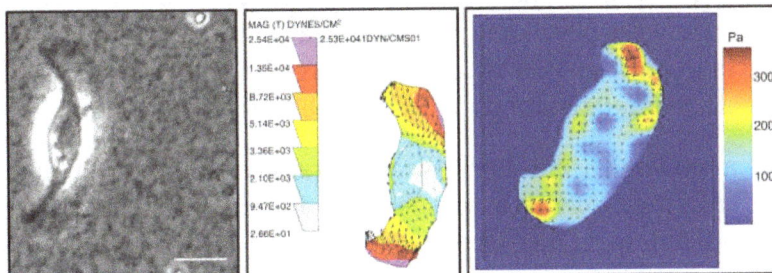

In figure, use of elastic substrates to map tractions in migrating cells. (A) Phase image showing a fish keratocytes crawling on an elastic polyacrylamide substrate. (B) Tractions mapped on the same cell shown in A. The Dembo Boundary element method algorithm was used to calculate the cell traction distribution from the bead displacement map; the units in the map are in Dynes/cm^2 (1 dyne = 10^{-5}N). (C) The Fourier-transform traction cytometry (FTTC) algorithm was used to calculate tractions for another keratocyte; the right scale of color bar represents stress in units of Pa (1 Pa = 1 N/m²). Scale bar is 10 μm.

However, with silicone rubber films, matching the compliance to the tractions exerted by the cells and providing a defined surface coating on the film for optimal adhesion is not always easy. These difficulties were circumvented by developing polyacrylamide gel substrates with variable degrees of cross-linking onto which ECM proteins could be conjugated. An example of the use of polyacrylamide substrates for examining the tractions exerted by locomoting keratocytes in seen in figure. Moreover, these films are optically tractable so that when fluorescent beads are used as the fiduciary markers in the gel, dual channel fluorescence microscopy permits the correlation of tractions in relation to the spatial localization of fluorescently labeled FA proteins.

Another approach employs special microfabricated substrates that contain an array of force-sensing elements. These are flexible cantilevers of known bending stiffness so that the forces exerted by moving cells on these pads can be computed directly from the deflection of the cantilever beams. An alternate approach employs an elastomeric silicone substrate that is micropatterned to give rise to a regular array of either surface indentations or projections of sub-micron dimensions. An algorithm allows the surface distortion of the micropattern caused by cells to be directly translated to the cellular forces. Thus, a number of methods now exist that are similar in overall concept and permit calculation of traction stresses and the correlation of those stresses with the molecular constituents of the force-transmitting adhesive structures.

Force, Cell Adhesions and Cell Migration

There is a clear interplay between contractile force generated by the cell, adhesion to the substrate and the traction applied to the substrate. As stated previously, force can induce focal complexes to mature into large FAs near the leading edge of migrating cells; at the trailing edge, contractile forces regulate adhesion disassembly and cell detachment. Also, MT-induced adhesion disassembly has been observed as mentioned previously and it was speculated that the growth of stiff MT growth into adhesions can release the force originally exerted by the actomyosin cytoskeleton, thus promoting adhesion disassembly.

The relationship among adhesion, traction applied to the substrate, and cell migration is under active investigation. At the outset, it is important to note that the net traction to move the cell through a low viscosity buffer is effectively zero. This leads to the conclusion that the typical tractions measured, which are much larger than what are required to move the cell, must be used to break adhesions in spatiotemporal patterns that dictate both the speed and direction of the cell.

Using keratocytes as a model, Lee and her colleagues reported that slowly migrating keratocytes are more fibroblast-like in their migration and characterized by slipping of adhesions that are coupled with retrograde actin flow; in fast-moving keratocytes, adhesions have more gripping character to sustain the rapid protrusion powered by the fast-paced polymerizing actin network; these cells exhibit a much smaller rearward actin flow. Recently, maps of actin-substrate coupling were used to quantify differences in force-transmission efficiency between different cell regions. Thus, a more detailed scenario about the substrate adhesion-traction-migration relationship could be proposed: At the leading edge, traction was transmitted in a manner partially independent of actin velocity (gripping) but at the cell flanks, the force transmission was mediated by the high friction between the actin network and the substrate; at the cell body, little traction was transmitted, because of low friction. Undoubtedly, this relationship will be further investigated both experimentally and theoretically, as it is key to achieving a global understanding of how cells move.

Collective Cell Migration

Collective cell migration is the prevalent mode of migration during development, wound healing, and tissue regeneration. In addition, it is increasingly regarded as a widespread mode of migration during metastasis in epithelial cancers. Collectively migrating cells use similar mechanisms as single cells to protrude, polarize, contract, and adhere to the surrounding matrix. However, their ability to interact with each other both chemically and mechanically provides cells within the moving group with additional mechanisms to migrate while (a) maintaining tissue cohesiveness and organization; (b) regulating tissue paracellular permeability; (c) creating large gradients of soluble factors; (d) distributing tasks between specialized mobile and non-mobile cells; (e) propagating mechanical signals via cell-cell junctions; and (f) in the case of cancer, protecting metastatic clusters from an immune assault. The mechanisms underlying collective cell migration are less well understood than those that drive single-cell migration, but improved methods in genomics, proteomics, imaging, and biomechanics are producing rapid advances in this field.

Collective Cell Migration in Physiology and Pathophysiology

The EMT paradigm

The transition from a static to a motile phenotype that multicellular collectives undergo during embryogenic movements, cancer metastasis, and wound repair is traditionally understood under the rubric of the epithelial to mesenchymal transition (EMT). EMT is a highly conserved cellular program characterized by a number of morphologic, structural, and molecular changes that includes flattening of cell shape, loss of apical-basolateral polarity and cell contacts, formation of a dynamic protrusion at the leading edge, and increased concentration of intermediate filaments. The EMT paradigm has been successful in explaining the transition between two well-defined cellular states as in the case of emigration of neural crest cells or gastrulation of mesoderm cells. However, many other processes involving cell motility do not follow the guidelines of classic EMT. This is illustrated by well-known cases such as migration of the zebrafish lateral line primordium (LLP) or that of border cells in *Drosophila* egg chamber. In both cases, cells efficiently migrate long distances while keeping a cohesive structure with high levels of E-cadherin and tight junction proteins. Thus, rather than being restricted to sharp transitions from well-defined epithelial and mesenchymal states, the cells in tissues can take advantage of both states to fine tune their phenotype. The spatial organization of this finely tuned phenotype within a moving group provides the cell with additional functions that could not be achieved in completely dissociated cell collectives.

Development

From early embryogenesis to postnatal life, the development of living organisms is driven by the motion of cell collectives. These collectives move in a variety of geometrical configurations such as sheets, sprouts, strands, tubes, or clusters. Given the difficulty of studying the motion of these geometrically diverse cell collectives within higher animals, collective cell migration is commonly studied in relatively simple models such as *Drosophila melanogaster*, *Caenorhabditis elegans*, or zebrafish. These model systems offer structural simplicity and genetic accessibility thus allowing the direct visualization of motile groups that selectively express fluorescently tagged proteins.

The simplest and best-studied mode of collective cell migration is the advance of 2D epithelial sheets over a basement membrane. Moving cell sheets are typically formed by a relatively large number of cells that remain mostly cohesive as they invade open spaces or surrounding tissues. A paradigmatic example of sheet migration is dorsal closure in *Drosophila*, a process occurring during the latest stages of embryogenesis. As a consequence of retraction of the germ band, an eye-shaped hole covered by amnioserosa cells is left on the dorsal surface. To seal this hole, two flanks of epidermal cells advance toward each other until they meet at the dorsal midline. This process is driven by at least three mechanisms: (1) active migration of epidermal cells characterized by dynamic extension of filopodia at the leading edge; (2) periodic contraction of the amnioserosa; and (3) zipping of supracellular actin cables at the leading edge. A different developmental process also driven both by supracellular purse-string and filopodia-rich migration is ventral enclosure in *C. elegans*. These examples illustrate that even in the simplest cases of collective migration, the motion of the group is not only driven by independent the action of each individual cell within the moving group but also by supracellular mechanisms and by the cooperative action of the surrounding tissue.

The inverse process to epithelial closure is centrifugal expansion of a cell sheet or colony. In some cases, such as the growth of imaginal discs in *Drosophila*, expansion occurs in the absence of physical constraints that restrict the growth of the colony. However, in other cases, cell colonies expand at the expense of the surrounding tissue. During development of the *Drosophila* abdomen, for example, the larval epithelium of each segment of the abdomen is replaced by proliferation and migration of four pairs of histoblast nests with original sizes ranging from 3 to 18 cells. These histoblast nests remain growth arrested during larval stages but at the onset of metamorphosis they undergo rapid proliferation and expand radially over each abdomen segment. Expansion occurs against the surrounding larval cells, which undergo apoptosis as they come in contact with the leading edge of the expanding histoblast colony. The mechanism by which histoblasts replace larval cells remains unknown but recent imaging improvements have shown leading histoblasts forming intercalating protrusions between the surrounding larval cells possibly contributing to their apoptosis. These protrusions are highly dynamics and suggestive of active traction generation.

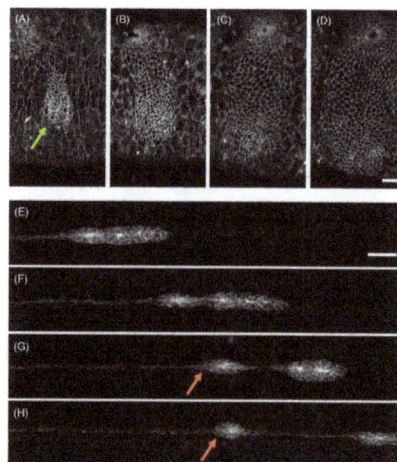

Collective cell migration in development

In figure, (A-D) During development of the abdomen of *Drosophila melanogaster* a cluster of histoblasts (green arrow) grows and migrates radially outward at the expense of the surrounding larval cells. Courtesy of Enrique Martin-Blanco and Carla Prat. (E–F) During development of the sensory system of zebrafish, the lateral line primordium undergoes directed migration from head to tail, leaving behind rosettes (red arrows) at periodic intervals. Scale bars: 60 μm.

In many processes in development, relatively small cell clusters undergo large collective displacements across the embryo from the location where they are specified to the location where they will ultimately carry out their biological function. A well-studied model for this type of cluster motion is the development of the lateral line system in zebrafish. The lateral line comprises a series of sensory organs that are arranged in regularly spaced clusters (neuromasts) on each flank of the skin. These clusters are deposited by the LLP, a group of 100 cells that migrates from head to tail. As the primordium transverses the animal, cell clusters at its trailing edge become progressively nonmotile and are finally left behind in periodic intervals. Another well-studied example of directed cluster motion is the migration of border cells during oogenesis in *Drosophila*. The border cell cluster comprises 6 to 8 migratory border cells and two nonmigratory inner polar cells. After being specified at the anterior end of the egg chamber, the cluster detaches from the follicle and migrates posteriorly toward the oocyte squeezing between the surrounding nurse cells. To do so, all border cells extend protrusions while they exchange positions within the cluster. Remarkably, border cells are not surrounded by ECM and, therefore, they need to adhere directly to nurse cells to generate traction.

Another developmental process in which collective cell migration plays a key role is branching morphogenesis, a process widely used in nature to shape complex organs that require packing of large surfaces into small volumes as in the case of lungs, kidneys, pancreas, mammary glands, salivary glands, and the vasculature. All these branched systems are characterized by the presence of a cell monolayer—epithelial, endothelial, or both—that separates two compartments within an organism. This cell layer controls the transport of ions, gases, liquid, solutes, and immune cells between these compartments. The formation of the initial branch commonly starts early in development by invagination of polarized epithelial cell sheets driven by constriction of the apical actomyosin rings. Alternatively, tubes can start by formation of a lumen from fusion of vacuoles or by sprouting of an already existing tube. Once the main branch is formed, emergence of new branches occurs either by splitting of one branch at its tip into two branches (bifurcation) or three branches (trifurcation), or by formation of a new branch from the side of an existing one (lateral branching). Collective migration and elongation of tubes appears to be led by extension of filopodia and lamellipodia from tip cells into the surrounding extracellular tissue. Such migration is regulated by the exchange of promigratory [EGF, fibroblast growth factor (FGF), vascular endothelial growth factor (VEGF), PDGF, etc.] and inhibitory factors (TGFβ, notch) between the epithelium and the mesenchyme. Interestingly, some of these molecular mechanisms are common not only in the morphogenesis of hollow organs but also in branching of the nervous system. In addition to these soluble factors, collective migration during branching morphogenesis is heavily regulated by the interaction between migrating cells and the ECM. Indeed, mutations in the cell-ECM adhesion proteins or in proteins involved in ECM degradation, as well as alterations in the composition of the ECM, result in reduced outgrowth of branches.

Wound Repair

The molecular machinery that governs collective cell migration during development remains largely dormant throughout adult life. However, when a tissue is injured, this machinery is rapidly rescued to repair the wound and restore the viable tissue. Wound healing plays a central role in the pathophysiology of virtually every organ. For example, in devastating lung diseases such as pulmonary fibrosis, chronic obstructive pulmonary disease, asthma, and acute lung injury, the pulmonary epithelium is injured and often denuded. This injury impairs barrier function of the epithelium thereby exposing the lungs to airborne inhaled pathogens and other toxic agents. In addition, the damaged airway epithelium prevents key metabolic functions of the airways including fluid and ion transport to the lumen and mucociliary clearance. The ability of the epithelium to rapidly self-repair is thus critical to restore pulmonary function and to prevent further damage.

Wound repair is particularly well understood in the skin but increasing evidence supports that the main stages of the process are conserved across organs. The initial physiological response to wounding is the activation of circulating platelets at the site of vascular injury. Such activation is initiated by direct contact between the platelet surface and proteins located at the basement membrane of the endothelium such as collagen, fibronectin, laminin, and von Willebrand factor. Activated platelets rapidly aggregate to form stable clots that prevent hemorrhage until the healing process is completed. Platelet aggregates are initially stabilized by a fibrin network that will later serve as a provisional scaffold rich in growth factors on which cells may crawl. In parallel with fibrin clotting, damaged cells initiate a stress response that includes the activation of MAPK pathways, the secretion of chemotactic factors, and the recruitment of circulating neutrophils and monocytes to clear pathogens from the injured area.

This initial inflammatory response is followed by reepithelialization. During this process, cells surrounding the wound migrate collectively across a provisional matrix rich in fibrin and fibronectin. To migrate onto and through this provisional matrix, cells at the first few rows behind the wound margin alter their expression of cell-cell and cell-matrix adhesion proteins. Fibrinolytic enzymes such as plasmin and MMPs degrade the matrix to enable rapid cell migration. In addition, cells undergo structural changes of their cytoskeleton characterized by the synthesis of transverse stress fibers and by the extension of filopodia and lamellipo-dia into the wound area. In striking analogy with development, epithelial cells use two main modes of collective migration during reepithelialization. The first mode involves the assembly of a supracellular actin cable at the wound perimeter. Contraction of this actin cable in a "purse string" manner provides efficient closure at the later stages of wound healing. The second mode of migration involves the extension of dynamic lamellipodia and filopodia into the wound area. This mechanism appears to be reminiscent of single cell migration although recent studies proved that it also involves strong cooperativity between cells.

In addition to playing a central role in reepithelialization, collective cell migration is also involved in wound healing as a primary mediator of angiogenesis. Angiogenesis is fundamental during wound healing to provide oxygen and nutrients to the newly assembled tissues and its inhibition impairs wound healing. It is mainly triggered by growth factors such as bFGF, TGFβ, VEGF, and by cytokines such as TNFα secreted by hypoxic macrophages and by damaged endothelial cells. In response to these signaling macromolecules, endothelial cells upregulate integrins at the tips of sprouting capillaries to collectively migrate through the surrounding tissue. As in the case of reepithelialization, proteolytic enzymes released into the wound tissue degrade the ECM to favor the advance of endothelial cell sprouts.

Cancer

While collective cell migration is crucial in development and tissue repair, it also mediates devastating diseases such as cancer. The traditional view of cancer metastasis is based on the notion that single cells detach from primary tumors, crawl through the stroma, enter the blood and lymphatic vessels, and finally colonize in healthy tissues to form a secondary tumor. However, increasing evidence indicates that tumor dissemination is driven not only by single cells but also by cohesive cell groups. This notion is supported by the observation that clusters of metastatic cells are often present in the blood and lymphatic vasculature of cancer patients. In addition, histopathological sections of breast, colon, ovarian, lung, and other differentiated carcinomas exhibit clusters, chains, and sheets in the stromal areas surrounding primary tumors.

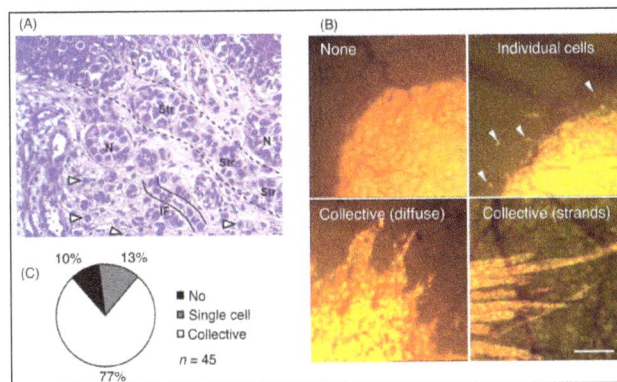

Collective cell migration in cancer

In figure, (A) Different invasion patterns in primary melanoma invading the mid-dermis *in vivo*. Arrowheads indicate scattered individual cells. Collective invasion modes include solid stands (Str), nests (N) representing cross-sectioned strands, and single cell chains (IF, "Indian files"). H&E staining. Image modified, with permission, from Friedl and Wolf. (B) Invasion modes in a modified skin-fold chamber model of orthotopic invasion of human HT-1080 fibrosarcoma cells. Patterns include lack of invasion (top, left), disseminating single cells (top, right), and diffuse or compact strand-like collective invasion (lower panels). Bar 250 μm. (C) Frequency of invasion modes displayed in B.

One successful strategy to study the role of these cohesive cell aggregates in cancer metastasis has been to analyze the dynamics of neoplastic tissue explants or cell line tumor spheroids *in vitro*. When embedded in 3D collagen I gels or Matrigel, these cell systems extend multicellular chains or strands into the surrounding matrix. Collective migration of this kind is initiated either by the polarization of a single cell within the cluster or by the activation of fibroblasts from the tumor stroma. These leading cells initiate the formation of a migration track by both cleaving and remodeling the surrounding matrix. The cooperative proteolytic activity of leading cells and their followers ultimately results in the generation of large invasive paths into the stroma.

Our mechanistic understanding of collective invasion in cancer is currently undergoing rapid progress thanks to the development of intravital microscopy. This technique enables the continuous monitoring of the dynamics of tumor tissue implanted in animal models. Typically, the implanted cells are fluorescently labeled with indicators of promoter activity, enzyme activity, or gene expression. Intravital microscopy has demonstrated the coexistence of single and collective cell invasion in a variety of organotypic cancer models. For example, implantation of HT-1080 fibro-sarcoma cells in dorsal skin-fold chambers of mice showed that up to 77% of invasive events were collective. Cells forming such invasive sheets or strands display heterogeneous phenotypes. While innermost cells in the clusters retain epithelial polarity and cell junctions, marginal cells display mesenchymal traits such as loss of apical-basalolateral polarity, actin-rich protrusions, and proteolytic activity. Recent studies using an organotypic model of breast mammary tumor metastasis showed that single-cell migration following dissemination from a primary tumor is relatively fast and capable of creating lung metastases via blood vessel circulation. By contrast, collective cell invasion is much slower and mainly invades lymph vessels. The switch from single- and collective cell invasion requires activation of a transcriptional program involving TGFβ and Smad4.

Mechanisms of Collective Cell Migration

Adhesion

To move as cohesive groups, cells require both cell-matrix and cell-cell adhesions. To a large extent, the molecular basis of cell-matrix adhesion in collective cell migration is analogous to that of single-cell migration.

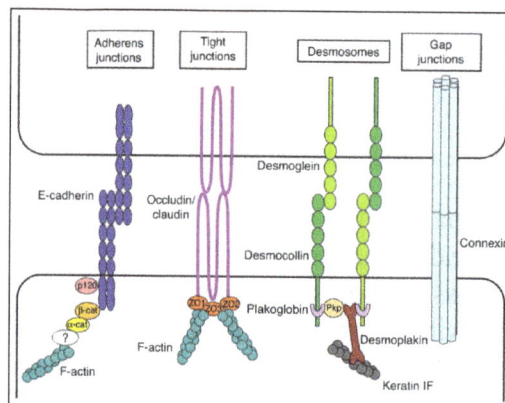

Scheme depicting the key molecules that mediate cell-cell adhesion during collective cell migration.

Adherens Junctions

Adherens junctions are responsible for a wide range of cellular functions including assembly and maintenance of cell-cell adhesions, stabilization of the actin cytoskeleton, and transcriptional regulation. Adherens junctions are based on the generally homophilic interaction between transmembrane glycoproteins of the calcium-dependent cadherin family. Currently more than 350 members of this family have been described in about 30 species, with epithelial (E-) cadherin being the best characterized. The extracellular domain of classical cadherins is composed of five domain repeats (EC1-EC5), which bind calcium ions to form parallel homodimers. Transpairing of the EC1 domains between cadherins from adjacent cells is required for proper conformational organization of adherens junctions, but other EC domains are likely to mediate cell-cell adhesion as well.

The cytoplasmic domain of cadherins is formed by two subdomains that mediate junctional stabilization and binding to the actin cytoskeleton. The subdomain that lies closer to the cell membrane is termed juxtamembrane domain (JMD). This domain contains a highly conserved octopeptide sequence that binds p120-catenin. The binding of p120-catenin and JMD is thought to retain cadherins at the plasma membrane thus providing stronger adhesion to the junction. In addition to stabilizing and strengthening junctions, p120-catenin also regulates single and collective cell motility via small GTPases. The cytoplasmic subdomain of cadherins that lies furthest from the cell membrane binds to β-catenin with high affinity in a phosphorylation-dependent manner. Until recently, β-catenin was thought to provide the physical link between cadherins and the actin cytoskeleton via α-catenin. This notion was supported by evidence showing that α-catenin is able to bind both actin and the β-catenin-cadherin complex. However, recent findings ruled out this possibility showing that β-catenin, α-catenin, and actin do not coexist in a single ternary complex. This finding raises the question of how adherens junctions are linked to the cytoskeleton. One possibility is that this link is mediated by an additional protein, such as EPLIN, an actin binding protein recently shown to bind the C-terminal domain of monomeric α-catenin.

Cohesiveness is usually associated with reduced migration speed. This was clearly illustrated in recent wound-scratch screens which showed that knocking down β-catenin and other key members of the adherens junction family in MCF-10A cells caused acceleration of cell migration. For this reason, whenever rapid migration occurs in nature, cells tend to downregulate E-cadherin and dissociate through a complete EMT. However, in many physiological situations, cells undergo an incomplete EMT in which E-cadherin adhesion is weakened to enable dynamic flexibility for each individual cell within the group while keeping a certain degree of cohesiveness. This weakening of adherens junctions is regulated by several signaling networks including those triggered by tyrosine kinases such as hepatocyte growth factor (HGF) receptor, epidermal growth factor receptor (EGFR), Eph receptor, Src, and Abl. These and other kinases regulate adherens junction strength by cadherin endocytosis, proteolysis, or interaction with other transmembrane proteins. Cadherin complexes regulate actin dynamics mainly via α-catenin, which inhibits Arp2/3-mediated branching polymerization and recruits the actin nucleator formin to adherens junctions. In addition to their role of providing junctional stability, β-catenin and p120-catenin can act as transcriptional regulators.

Tight Junctions

Tight junctions are located apically from adherens junctions both in static monolayers and in migrating epithelia. They are thought to play a double role: first, they serve as the intramembranous "fences" that separate the protein content of the apical and basolateral cell membranes; second, they are hydrophobic "barriers" that regulate transport of ions, proteins, and fluids across epithelial and endothelial layers.

Tight junctions comprise three main types of transmembrane proteins: occludins, claudins, and the IgG-like family of junctional adhesion molecules (JAMs). Occludin is a tetra-spanning-transmembrane protein with two extracellular loops that can be phosphorylated at multiple tyrosine, serine, and theonine residues. In the absence of phosphorylation, ocludins are localized throughout the basolateral cell membrane and in cytoplasmic vesicles, but phosphorylated occludins are only present at tight junctions. Recently, a new protein with a similar structure and role as occludin, tricellulin was found to be enriched only at tricellular tight junctions. Despite their ubiquitous presence in tight junctions, a number of studies demonstrated that cells and tissues deficient for occludin display proper barrier function. This observation led to the identification of claudins.

Much like occludins, claudins are also tetra-span-transmembrane proteins with two extracellular loops. The claudin family comprises at least 24 members that are specifically distributed across organs and tissues. This distribution selectively tunes the size, strength, and transport specificity of the junctions. In addition to tetra-span proteins, tight junctions also contain single spanning transmembrane proteins that mediate homotypic adhesion. These proteins include the IgG-like JAMs, which mediate paracellular transmigration of leucocytes.

Given that occludins, claudins, and JAMs have not been found to interact directly, it is thought that the integrity of tight junctions is mediated by scaffolding proteins such as ZO-1, ZO-2, and ZO-3. These proteins interact with claudins and occludins though their PDZ domains, and with actin through their C-terminus thus providing a direct connection between the extracellular environment and the cytoskeleton. ZO-1 can bind actin and α-catenin, which has led to the hypothesis that ZO-1 may serve as a link between adherens junctions and tight junctions.

Tight junctions are thought to play a central role in finely tuning apical-basolateral polarity within moving groups but mechanism remains poorly understood. A wealth of evidence supports that preservation of intact tight junctions prevents tumor dissemination by inhibiting cell proliferation and migration. However, several studies have demonstrated the existence of epithelial polarity among invasive tumors suggesting that tight junctions remain functional during certain modes of invasion. Tight junction proteins have also been reported to contribute to enhanced invasion and collective cell migration. For example, ZO-1 was found to be upregulated in a high proportion of highly metastatic melanoma cell lines. Similarly, overexpression of claudin-3 and claudin-4 in human ovarian epithelial cells resulted in increased collective migration in wound healing experiments. Tight junctions also play a central role during collective cell migration in various developmental processes. In *Drosophila*, mutations in the ZO-1 homologues result in defects in tracheal morphogenesis and in the formation of extrasensory organs. In zebrafish, the posterior LLP elicits a homogeneous distribution of cadherin among all cell-cell junctions, but ZO-1 is absent in the first few rows behind the leading edge. However, toward the trailing edge of the migrating primordium, the emergent proneuromast rosettes display an apical formation of tight junctions before being deposited. These findings indicate that within a cohesive moving group tight junctions can be selectively formed to control apical-basolateral polarity.

Desmosomes

Desmosomes are intercellular junctions that connect the intermediate filaments from adjacent cells. They are commonly found in tissues that are subjected to substantial mechanical forces such as the epithelia and muscle. Extracellularly, desmosomes are similar to adherens junctions in that extracellular linkers are transmembrane proteins with five domain repeats homologous to classical cadherins. A functional desmosome contains at least one desmosomal cadherin from the desmocollin family and another one from the desmoglein family. The cytoplasmic domains of desmosomal cadherins interact with the armadillo proteins, plakoglobin (gamma-catenin), and plakophilins. The role of plakoglobin in its binding to desmosomal cadherins is reminiscent to that of β-catenin in adherens junctions. Plakoglobin binds to the N-terminal of desmoplakin, the protein that ultimately binds desmosomes to intermediate filaments. The role of plakophilin is more complex than that of plakoglobins as it involves interactions with desmosomal cadherins, plakoglobins, and intermediate filaments. These complex cytoplasmic interactions constitute the molecular clustering that provides strength to desmosomes.

Several lines of evidence indicate that desmosomal adhesion is regulated during collective cell migration. Upon wounding Madin-Darby canine kidney (MDCK) monolayers, desmosomes switch from a Ca2+-independent to a Ca2+-dependent phenotype. This loss of Ca2+ independence is followed by desmosome internalization mediated by protein kinase C (PKC). The resulting loss of adhesion provides the epithelium with the flexibility required to achieve fast reepithelialization. It remains unclear, however, if the loss of desmosomal adhesion is complete or partial. In colorectal tumors, two types of desmocollins (Dsc1 and Dsc 2) are expressed *de novo*, suggesting collective cell invasion. In squamous cell carcinomas of the skin, desmoglein 2 is upregulated and the levels of desmoglein activation correlate with risk of metastasis. These findings indicate that desmosomes might remain functional during tumor invasion in some forms of cancer.

Gap Junctions

Gap junctions are transmembrane channels that connect the cytoplasm of adjacent cells. Each cell contributes to the junction with half a channel (an hemichannel or connexon) formed by six proteins, termed connexins. Each connexin comprises four transmembrane domains connected by two extracellular loops that mediate cell-cell recognition and intercellular docking. Connexins are arranged in a cylindrical patterns that leave a hollow central channel for transit of ions, second messengers, and small metabolites with molecular weights lower than 1 kDa.

As central mediators for cell-cell communication, gap junctions have long been implicated in the regulation of collective cell migration during cancer metastasis, wound healing, and morphogenesis. Tumor cells from several human cancers such as skin, lung, gastric, and prostate cancers, exhibit reduced expression of gap junction proteins and reexpression of these proteins appears to play a tumorsuppressive role. For example, overexpression of Cx26 has been shown to slow down collective migration of the breast cancer cell line MCF-7 and to reverse its malignant phenotype. Similarly, exogenous expression of Cx26 and Cx43 also reduced migration of MDA-MB-231 and MDA-MB-435 breast cancer cells. Contrary to this widespread notion, recent evidence demonstrates that during certain stages of metastasis Cx26 is reexpressed suggesting cell cooperativity during invasion. These findings indicate that gap junctions might selectively regulate the transition from single- to collective cell migration during different stages of cancer progression.

Further support for gap junction activity during collective cell migration comes from wound-healing experiments. During epidermal wound healing, the expression of connexins is altered as a function of the distance from the wound. Specifically, Cx26 was found to be downregulated at the wound edge, but upregulated away from the wound. By contrast, Cx31.1 and Cx43 were downregulated both at the wound edge and further away from the wound edge. In an *in vitro* wound-scratch assay using MCF-10A cells, knocking down Cx43 accelerated collective cell migration, whereas silencing Cx26 and Cx40 had no net effect. Perhaps the clearest illustration of the pivotal role of gap junctions in regulating collective cell migration can be found in development, where mutations or silencing of genes associated with gap junctions result in abnormal migration. For example, mutations of innexin, the connexin homolog expressed in invertebrates, prevent collective epithelial cell migration during proventriculus organogenesis in *Drosophila*. In mouse models, the levels of expression of Cx43 correlate positively with speed and directionality of neural crest migration. Taken together, the studies mentioned previously support the notion that gap junctions play a central role in the regulation of collective cell migration, but the molecular mechanisms underlying this regulation and the interaction between gap junctions and other cell-cell junctions remain largely unknown.

Polarization and Guidance

One of the great advantages of collective versus single-cell migration is that each cell within the moving group can exhibit different patterns of expression to carry out specialized functions according to its position within the group. The simplest case of such modular specialization is front-rear polarization in which one subset of leader cells at the front guides a larger group of naïve follower cells at the rear. Polarization of this kind can arise as a consequence of internal genetic programs or external environmental cues. Leader cells typically exhibit a mesenchymal-like phenotype characterized by the extension of lamellipodia and filopodia into the surrounding tissue, a relatively loose cell-cell adhesion, enhanced expression of cell-matrix adhesion proteins, and polarized remodeling of actin filaments and MTs In addition, leader cells moving in 3D are capable of degrading and remodeling the ECM to create channels for the whole cell group to advance cohesively. By contrast, followers retain epithelial features such as apical-basolateral polarity and tight junctions and express relatively low levels of guidance receptors.

A clear example of leader/follower polarization occurs during angiogenesis. Tip (leader) cells extend numerous filopodia that probe, guide, and presumably generate tractions to drive motion of the tubes to the avascular area of the embryo. In contrast to their follower stalk cells, tip cells are nonlumenized and mostly nonproliferative. In addition, they exhibit a clearly distinct pattern of gene expression with higher levels of expression of VEGF receptor family, which tightly controls the generation of sprouts during angiogenesis. While gradients of VEGF guide tip cells, the concentration of VEGF regulates proliferation in stalk cells. A central question in angiogenesis is how the tip cell is initially selected from a large pool of endothelial cells exposed to similar VEGF gradients. In other words, why do endothelial cells form sprouts and branches instead of sheets? The answer to these questions lies in the competitive advantage that tip cells gain by signaling their neighbors to become stalk cells. This is accomplished via the Notch signaling pathway. VEGFR2 activation causes upregulation and release of the Notch ligand Dll4 at the tip cells. The resulting activation of Notch in the neighboring cells leads to downregulation of VEGFR2 and ultimately the acquisition of stalk phenotype.

Another example of collective cell migration guided by a chemoattractant gradient is border cell migration. In these cells, chemoattractant gradients are sensed by two receptors, EGFR and poliovirus receptor (PVR), each of which can independently guide cell migration. During the early phase of border cell migration, these receptors act at the sub-cellular level to drive polarization and guide migration much as in the case of single isolated cells. Each cell individually senses the gradient and acts accordingly resulting in highly persistent directional migration of the cluster. In a later phase, the very same receptors and chemoattractant cues appear to act at a higher level of organization in which the intercellular differences in the levels of signaling from the guidance receptors cells determines the identity of the front cell. In this case, cells within the cluster compete to guide the group thus constantly exchanging positions throughout the collective process. This guidance strategy results in slower overall motion, but it offers a broader range of possibilities to probe the cluster environment.

Cell groups can also migrate in the absence of chemoattractant gradients by using inherent front/rear polarization. To migrate from the anterior to the posterior regions of the embryo, the posterior LLP follows a track defined by a strip of the chemokine stromal-derived factor, SDF1a Although the precise values of the concentration of SDF1a along the strip remain unknown, it appears that SDF1a is uniformly distributed. This notion is supported by the observation that the primordium is able to perform a U-turn and migrate backward. To perform directed migration in the absence of a chemoattractant gradient, the primordium acquires front/rear polarity through the localized activation of two SDF1a receptors, CXCR4 and CXCR7. Both receptors are required for migration of primordium but CXCR4 is active at the leading edge whereas CXCR7 is localized at the trailing edge. Polarization of these receptors is maintained through the interaction of the Wnt/β-catenin, Fgf, and Dkk pathways. In addition, recent evidence suggests that sequestration of SDF1a by CXCR7 might be a crucial event to determine the persistence of primordium migration.

The zebrafish LLP is also a representative model to illustrate that some cell collectives are able to achieve supracellular tissue patterning as they migrate. Roughly behind the leading third of the primordium, a group of 12 to 16 cells organizes in rosettes to form the proneuromasts. Cells within these rosettes display a marked epithelial phenotype characterized by columnar morphology, apical-basolateral polarization, and the presence of foci of the tight junction protein ZO1. After their formation, rosettes become progressively less motile until they are left behind and deposited at periodic length intervals. Thus the LLP illustrates that moving cell groups may display different levels of collective organization beyond front/rear polarity.

Mechanics of Collective Cell Migration

Kinematic observations and Models

The kinematics of collective cell migration has been the subject of experimental and theoretical investigation virtually since light microscopy was first applied to the observation of biological processes. Indeed, the first observations of tumor dissemination, growth of epithelial tissues, and wound closure date back more than one century. The advent of modern imaging techniques such as confocal microscopy, multiphoton microscopy, and intravital imaging, together with the development of improved fluorescent probes and computational methods now enable us to quantitatively analyze the kinematics of collective cell migration *in vivo*. Outstanding advances in this field include visualization of cancer cells within a collective cluster as they escape a tumor to invade lymphatic vessels, or tracking of hundreds of individual cells involved in mesoderm migration during development of the *Drosophila* embryo.

The study of cell kinematics combined with a variety of continuum and discrete physical models has provided substantial advances in our understanding of how cells move collectively. Typical continuum models are governed by reaction-diffusion equations:

$$\frac{\partial n}{dt} = D_n \nabla^2 c - \nabla . \left(\chi(c) n \nabla c \right) + f(n,c)$$

$$\frac{\partial c}{dt} = D_n \nabla^2 c + g(n,c)$$

Where, n is cell density and c is the concentration of chemotactic signal. The first term on the right hand side of $\dfrac{\partial n}{dt} = D_n \nabla^2 c - \nabla.\left(\chi(c)n\nabla c\right) + f(n,c)$ accounts for random cell migration and the second term models chemotaxis (or haptotaxis). In chemotaxis, for example, the chemoattractant is expressed or bound to a matrix, χ is the chemotactic sensitivity, which may be function of c, and the function f describes mitotic generation and natural cell loss. Conversely, the first term on the right hand side of $\dfrac{\partial c}{dt} = D_n \nabla^2 c + g(n,c)$ accounts for diffusion of the tactic agent, and $g(n,c)$ captures its production, uptake, and degradation. This system of reaction diffusion equations can be generalized to any number of cell types and tactic agents present in the system under investigation. Further coupling between equations can be obtained by taking the diffusion coefficient as a function of cell concentration.

Continuum models are limited by their relative inability to take into account dynamics of cell adhesion as well as local variability of the cell mechanical properties. These factors are introduced in the so-called vertex models or cellular Potts models, in which cell geometry is discretized and different mechanical and adhesive properties, can be associated with each constitutive element of the system. In these models, the cell collective explores an energy landscape determined by a 3-term Hamiltonian (H):

$$H = H_{adhesion} + H_{deformation} + H_{chemotaxis}$$

The first term on the right-hand side, $H_{adhesion}$, embodies the adhesive energy associated with cell-cell and cell-matrix interactions, the second term, $H_{deformation}$, accounts for the deviation of cell shape from a "preferred" geometry, and the third term, $H_{chemotaxis}$, models tendency of each cell to move toward maximum concentrations of chemotactic agents. For a given Hamiltonian, the dynamics of the system are obtained by energy minimization using Monte-Carlo strategies. Each element in the system is sampled in a random manner and given a new configuration. If this new configuration decreases the energy of the system the change is always accepted, otherwise it accepted with a given probability. This approach can be coupled to continuum reaction-diffusion equations to take into account the dynamics of diffusive chemicals associated with chemotaxis.

The modeling approaches described previously have been successful in reproducing the kinematics of collective cell migration in a variety of biological processes including tumor angiogenesis, wound healing, and cell sorting. But even if the kinetics is captured, the underlying mechanisms remain far from being elucidated. Consider, for example, the relatively simple case of 2D wound-scratch assays. The kinematics of this process has been captured by continuum and discrete models in which the key ingredient is the establishment of a chemotactic gradient. Thus, models of collective cell migration need to formulate further experimentally testable predictions to elucidate the relative contribution of both the biochemical and the biophysical mechanisms that drive collective cell migration.

Dynamics

Our relatively poor understanding of the mechanisms underlying collective cell migration is not so much due to the lack of suitable physical models as it is due to the lack of key experimental information. Perhaps the most important piece of experimental information we are lacking is a physical picture of the forces that drive collective cell migration. Without this information, it is not possible to determine whether collective cell motion is a local process in which each cell is mechanically self-propelled or an integrated process in which physical forces are transmitted across long distances to coordinate the action of each individual cell within a moving group.

The technique that has provided most information about the dynamics of living tissues *in vivo* is laser microsurgery. This technique is based on the selective ablation of single cells within tissues. The analysis of the resulting tissue relaxation enables the inference of the actual state of stress of the tissue. By applying this technique to dorsal closure in *Drosophila* embryos, Kiehart et al. showed that both the leading edge of the lateral epidermis and the amnioserosa are under tension. In a later study, Hutson et al. used straightforward force balance at the leading edge to conclude that the tensile contribution of amnioserosa and the epidermal tissues is of similar magnitude while the contribution of actin ring contraction is 2-fold higher. Other findings obtained by laser ablation include the measurement of contractile forces associated with cell apoptosis, and the guidance of tissue morphogenesis by anisotropic forces.

While laser ablation methods enable the inference of the state of stress of tissues as well as their dynamic relaxation, they do not provide maps of the forces associated with cell migration. The first such maps were obtained by the joint of effort of the groups of B. Ladoux and P. Silberzan. The authors seeded epithelial cell colonies on top of a micropillar array and observed the time evolution of the forces exerted by the cells on the pillars. They showed that forces at the leading edge are tensile, thus ruling out that the epithelial tissue advances as a result of pushing forces from submarginal cells. In addition, the authors remarked that submarginal cells are also able to generate traction forces of substantial magnitude, which is consistent with the observation by Farooqui and Fenteany that submarginal cells extend cryptic lamellipodia under their neighbors at the margin. A later study by Trepat et al. showed that not only submarginal cells are able to generate substantial traction forces, but also that these forces are integrated over long distances to generate a stress gradient at cell-cell junctions. The existence of such stress gradients combined with mechanotransduction events at cell-cell junctions might provide novel mechanisms of positional sensing within moving groups.

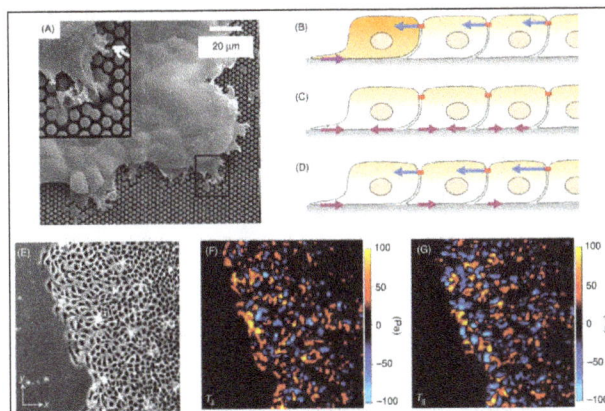

Mechanics of collective cell migration

In figure, (A) The forces exerted by the leading edge of an MDCK epithelial cell sheet migrating on top of a microneedle array are tensile. Adapted, with permission, from reference. (B, C, D) Patterns of force generation and transmission in an epithelial cell sheet. (B) An active leader cell generates forces at the leading edge and transmits these forces to follower cells via cell-cell junctions. (C) Each cell within the monolayer generates its own contractile forces. Forces are balanced locally in such a way that there is no force transmission through cell-cell junctions. (D) Tug-of-war force generation and transmission. The local tractions that each cell generates are transmitted through cell-cell junctions to generate a global gradient of tensile stress. (E) Phase contrast image of an MDCK cell sheet advancing on top of a soft polyacrylmide gel (1.2 kPa). In this model, tractions parallel (F) and perpendicular (G) to the leading edge rule out the existence of leader/follower polarity.

Recently, force microscopy technology was improved to enable the measurement of inter- and intracellular forces. Starting with tractions at cell-substrate interface and using straightforward force-balance imposed by Newton's laws, Tambe et al. developed monolayer stress microscopy (MSM) to map the state of stress at any point within a monolayer. Using this technology the authors showed that intracellular stresses vary abruptly across a migrating monolayer sheet, and that force transmission through cell-cell junctions expands several cell diameters. In addition, the authors showed in a variety of cell types that cell collectives move along the direction of maximum normal stress—or, equivalently, minimum shear stress. This mode of collective guidance was called plithotaxis.

Cytoplasmic Streaming

Cytoplasmic streaming, also called protoplasmic streaming is the movement of the fluid substance (cytoplasm) within a plant or animal cell. The motion transports nutrients, proteins, and organelles within cells. First discovered in the 1830s, the presence of cytoplasmic streaming helped convince biologists that cells were the fundamental units of life.

Although the mechanism of cytoplasmic streaming is not completely understood, it is thought to be mediated by "motor" proteins—molecules made up of two proteins that use adenosine triphosphate (ATP) to move one protein in relation to the other. If one of the proteins remains fixed on a substrate, such as a microfilament or a microtubule, the motor proteins can move organelles and other molecules through the cytoplasm. Motor proteins often consist of actin filaments, long protein fibres aligned in rows parallel to the streaming just inside the cell membrane. Myosin molecules attached to cellular organelles move along the actin fibres, towing the organelles and sweeping other cytoplasmic contents in the same direction.

Cellular Senescence

Cellular senescence is a process that results from a variety of stresses and leads to a state of irreversible growth arrest. Senescent cells accumulate during aging and have been implicated in promoting a variety of age-related diseases. Cellular senescence may play an important role in tumor suppression, wound healing, and protection against tissue fibrosis; however, accumulating evidence that senescent cell may have harmful effects in vivo and may contribute to tissue remodeling, organismal aging, and many age-related diseases also exists. Cellular senescence can be induced by various intrinsic and extrinsic factors. The pathways for the proteins p53/p21 and p16^{Ink4a}/retinoblastoma protein are important for irreversible growth arrest and senescent cells. Senescent cells secrete numerous biologically active factors; the specific secretion phenotype by senescent cell contributes to physiological and pathological consequences in organisms.

Cellular senescence refers to the essentially irreversible arrest of cell proliferation (growth) that occurs when cells experience potentially oncogenic stress. The permanence of the senescence growth arrest enforces the idea that senescence response evolved at least in part to suppress the development of cancer. A senescence arrest is considered irreversible because no known physiologic stimuli can stimulate senescence cells to re-enter the cell cycle. The senescence arrest is stringent; it is established and maintained by at least 2 major tumor suppressor pathways: the proteins p53/p21 and the p16^{Ink4a}/retinoblastoma protein (pRb). These pathways are now recognized as a formidable barrier to malignant tumorigenesis.

In addition to arrested growth, senescence cells show widespread changes in chromatin organization and gene expression. These changes include the secretion of numerous proinflammatory cytokines, chemokines, growth factors, and proteases, which is a featured function of senescence-associated secretory phenotypes (SASPs) he remaining causes of cellular senescence are beyond the scoThe SASP has powerful paracrine activities, the nature of which suggests that senescence response is not solely a mechanism for preventing cancer. Rather, cellular senescence and the SASP likely evolved to suppress the development of cancer and to promote tissue repair or regeneration in the face of injury.

Through the SASP, a low absolute number of senescent cells in a tissue (typically < 20%) may be able to exert systemic effects. For example, obesity-associated senescent cells may promote chronic, low-grade, sterile inflammation. In this way, senescent cells might be a link between obesity and inflammation that contributes to the development and progression of type II diabetes. Although cellular senescence is normally a defense mechanism against tumor development, the presence or persistence of a high number of senescent cells can promote tumor progression because of inflammation, tissue disruption, and growth signals due to the SASP. Senescent cells can also initiate a deleterious positive feedback mechanism by promoting the spread of senescence to nearby cells.

Senescent cell burden is low in young individuals but increases with aging in several tissues, including adipose, skeletal muscle, kidney, and skin. In particular, components of the metabolic syndrome, including abdominal obesity, diabetes, hypertension, and atherosclerosis, are among the many pathologies that are associated with increased senescent cell burden. Senescent cell accumulation can occur due to a variety of factors such as various age-related chronic diseases, oxidative stress, hormonal milieu, developmental factors, chronic infection (e.g., human immunodeficiency virus [HIV]), certain medications (chemotherapy or certain HIV protease inhibitors), and radiation exposure.

There are different types of cellular senescence that have been identified, including oncogene-induced senescence, stress-induced premature senescence as seen in patients with diabetes, and the classical replicative senescence. Therefore, senescent cells can contribute to aging and all age-related pathologies by accelerating loss of tissue regeneration through the depletion of stem cells and progenitor cells. Cellular senescence is indicated in every pathological condition associated with aging.

Causes

Cellular senescence was formally described by Hayflick in the 1960s. He showed that after undergoing a certain number of divisions, normal human diploid fibroblasts enter an irreversible nondividing state or replicative senescence. Research has demonstrated that normal human diploid fibroblasts can divide 50 to 60 times, but afterwards they stop dividing irreversibly. Thus, the number of divisions cells complete before reaching the end of their replicative lifespan has been termed as the Hayflick Limit.

Senescence has been reported to occur in a number of other cell types such as keratinocytes, melanocytes, endothelial cells, epithelial cells, glial cells, adrenocortical cells, T lymphocytes, and even tissue stem cells.

Even though senescence is induced by multiple factors such as repeated cell culture, telomere attrition, irradiation, oncogene activation, and oxidative damage, it can also be caused by the perturbation of mitochondrial homeostasis, which may accelerate age-related phenotypes. Because mitochondria can generate reactive oxygen species (ROS), it is proposed that excessive mitochondrial ROS are important to establishing cellular senescence. Perturbations of mitochondrial homeostasis will include excessive ROS production, impaired mitochondrial dynamics, electron transport chain defects, bioenergetic imbalances or increased adenosine monophosphate-activated protein kinase activity, decreased mitochondrial nicotinamide adenine dinucleotide or altered metabolism, and mitochondrial calcium accumulation.

There are several causes that can induce cellular senescence, which can also include telomere shortening, genomic damage, strong mitogen-associated signals, epigenomic damage, and activation of tumor suppressors. Replicative senescence is indeed not dependent on chronological time and culture, but rather depends on the number of divisions that cells undergo in culture. It is thought that telomere shortening, which occurs at each cell division because of incomplete replication, is the counting mechanism for the induction of replicative senescence. Functional telomeres prevent DNA repair machineries from recognizing chromosome ends as DNA double-stranded breaks, which the cells rapidly respond to and attempt to repair. In the case of telomeres, repair followed by cell division will cause rampant genomic instability through cycles of chromosome fusion and breakage, which are major risk factors for developing cancer. Therefore, repeated and extensive divisions in the absence of telomerase eventually cause 1 or more telomeres to become critically short and dysfunctional. Dysfunctional telomeres elicit a DNA damage response (DDR) but suppress attempted DNA repair. In turn, this DDR arrests cell division primarily through activities of the p53 tumor suppressor, thereby preventing genomic instability. Telomeres become critically short after extensive division, and telomere ends are recognized as DNA double-strand breaks. The telomere ends aggravate a DDR in cell divisions and are then arrested by the activated DDR, mainly through p53 tumor suppressor activity.

The mechanism behind the finite replicative lifespan of normal cells is now understood quite well. Because polymerase that copy DNA templates are unidirectional and require a labile primer, the ends of linear DNA molecules cannot be completely replicated. Thus, telomeres, the DNA protein structures that cap the ends of linear chromosomes, shorten with each cell division. Telomere shortening does not occur in cells that express telomerase, the reverse transcriptase that can replenish the repetitive de novo telomeric DNA.

The numbers and types of telomerase-expressing cells vary widely among species. In humans, however, such cells are rare. Telomerase-positive human cells include most cancer cells, embryonic stem cells, certain adult stem cells, and a few somatic cells (e.g., activated T cells).

Dysfunctional telomeres appear to be irreparable; consequently, cells with such telomeres experience persistent DDR signaling and p53 activation, which enforce the senescence growth arrest. The signaling for DDR also establishes and maintains the SASP. Therefore, when there are pathologies that cause DNA damage enough to stimulate and prevent repair of DNA, this will stimulate cellular senescence significantly.

Senescence-associated Secretory Phenotype

An important feature of many senescent cells is the SASP. The SASP is arguably the most striking feature of senescent cells, because it has the potential to explain the role of cellular senescence in organismal aging and age-related pathologies. Consistent with its complexity, the SASP biological activities are myriad. The SASP can stimulate cell proliferation, owing to proteins such as growth-related oncogenes and amphiregulin, as well as stimulate new blood vessel formation due to proteins such as vascular endothelial growth factor. However, the SASP can also include proteins that have complex effects on cells. For example, the biphasic WNT modulator secreted frizzled-related protein 1 and interleukin (IL)-6 and IL-8, which can stimulate or inhibit WNT signaling cell proliferation, depend on the physiological context.

Chronic WNT signaling can drive both differentiated cells and stem cells into senescence. Also, some SASP factors induce an epithelial-to-mesenchymal transition in susceptible cells. Thus, these aforementioned SASP factors can alter stem cell proliferation and differentiation or modify stem cell niches. In addition, many SASP components directly or indirectly promote inflammation, which is of particular importance to the role of cellular senescence in aging and age-related disease. These factors include IL-6 and IL-8, a variety of monocyte chemoattractant proteins and macrophage inflammatory proteins, and proteins that regulate multiple aspects of inflammation such as granulocyte-macrophage colony-stimulating factor. The secretion of these and similar proteins by senescent cells is predicted to cause chronic inflammation, at least locally and possibly systemically.

Chronic inflammation, of course, is a cause of or an important contributor to virtually every major age-related disease, both degenerative and hyperplastic. The SASP is also a plastic phenotype; this means proteins that are included in the SASP vary among cell types and, to some extent, with the stimulus that induced the senescence response. Nevertheless, there is substantial overlap among SASPs; pro-inflammatory cytokines are the most highly conserved feature, cutting across many different cell types and senescence-inducing stimuli.

Senescent Cells and Degenerative Phenotypes

Senescent cells have been implicated in many age-associated degenerative phenotypes, both normal and pathological. In most cases, senescent cells have been shown to drive degenerative changes, largely through their secreted proteins from their SASP. Senescent cells can disrupt normal tissue structures, which are essential for normal tissue function. Senescent cells and SASPs can also fuel overt age-related diseases. For example, indirect evidence shows that senescence and associated SASPs of astrocytes can promote the age-related neurodegeneration that gives rise to cognitive impairment as well as to Alzheimer's and Parkinson's diseases. In addition, the presence of SASP in senescent chondrocytes, which are prominent in age-related osteophytic joints and degenerated intervertebral discs, are thought to play a major role in etiology and promotion of these pathologies.

Moreover, senescent epithelial, endothelial, and smooth muscle cells have been implicated in the genesis and promotion of age-related cardiovascular disease. The list of age-related pathologies in which senescent cells have been observed and proposed to cause or contribute is long and includes macular degeneration, chronic obstructive pulmonary disease, emphysema, and insulin insensitivity, among others. Therefore, senescent cells are a smoking gun present at the right time and place to drive these age-related pathologies.

Cell Death

Cell death terminates normal cellular functions, including respiration, metabolism, growth and proliferation. Cell death can be non-programmed, for example as the result of accidental injury or trauma, or programmed.

Cell death can be classified according to its morphological appearance (which may be apoptotic, necrotic, autophagic or associated with mitosis), enzymological criteria (with and without the involvement of nucleases or of distinct classes of proteases, such as caspases, calpains, cathepsins and transglutaminases), functional aspects (programmed or accidental, physiological or pathological) or immunological characteristics (immunogenic or non-immunogenic).

Dying cells are engaged in a process that is reversible until a first irreversible phase or 'point-of-no-return' is trespassed. It has been proposed that this step could be represented by massive caspase activation, loss of $\Delta\Psi_m$, complete permeabilization of the mitochondrial outer membrane or exposure of phosphatidylserine (PS) residues that emit 'eat me' signals for normal neighboring cells. However, there are dozens of examples in which caspases are activated in the context of non-lethal processes and differentiation pathways. The $\Delta\Psi_m$ can be dissipated by protonophores without progression to immediate cell death. PS exposure can be reversible, for instance in neutrophilic granulocytes. Thus, the concept of a restriction point for cell death, as it was described by Pardee for the cell cycle, has yet to be specifically defined.

Table: Cell death methodology.

Definition	Notes	Methods of Detection[3-5]
Molecular or morphological criteria to define dead cells.		
Loss of plasma membrane integrity	Plasma membrane has broken down, resulting in the loss of cell's identity.	(IF) Microscopy and/or FACS to assess the exclusion of vital dyes, *in vitro*.
Cell fragmentation	The cell (including its nucleus) has undergone complete fragmentation into discrete bodies (usually referred to as apoptotic bodies).	(IF) Microscopy FACS quantification of hypodiploid events (sub-G_1 peak).

Engulfment by adjacent cells	The corpse or its fragments have been phagocytosed by neighboring cells.	(IF) Microscopy FACS colocalization studies.
Proposed points-of-no return to define dying cells		
Massive activation of caspases	Caspases execute the classic apoptotic program, yet in several instances, caspase-independent death occurs. Moreover, caspases are involved in non-lethal processes including differentiation and activation of cells.	Immunoblotting FACS quantification by means of fluorogenic substrates or specific antibodies.
$\Delta\Psi_m$ dissipation	Protracted $\Delta\Psi_m$ loss usually precedes MMP and cell death; however, transient dissipation is not always a lethal event.	FACS quantification with $\Delta\Psi_m$-sensitive probes Calcein-cobalt technique.
MMP	Complete MMP results in the liberation of lethal catabolic enzymes or activators of such enzymes. Nonetheless, partial permeabilization may not necessarily lead to cell death.	IF colocalization studies Immunoblotting after subcellular fractionation.
PS exposure	PS exposure on the outer leaflet of the plasma membrane often is an early event of apoptosis, but may be reversible. PS exposure occurs also in T-cell activation, without cell death.	FACS quantification of Annexin V binding.
Operative definition of cell death, in particular in cancer research		
Loss of clonogenic survival	This method does not distinguish cell death from long-lasting or irreversible cell cycle arrest.	Clonogenic assays

Abbreviations: $\Delta\Psi_m$, mitochondrial transmembrane permeabilization; FACS, fluorescence-activated cell sorter; IF, immunofluorescence; MMP, mitochondrial membrane permeabilization; PS, phosphatidylserine.

In the absence of a clearly defined biochemical event that can be considered as the point-of-no-return, the NCCD proposes that a cell should be considered dead when any one of the following molecular or morphological criteria is met: (1) the cell has lost the integrity of its plasma membrane, as defined by the incorporation of vital dyes (e.g., PI) in vitro; (2) the cell, including its nucleus, has undergone complete fragmentation into discrete bodies (which are frequently referred to as 'apoptotic bodies'); and (3) its corpse (or its fragments) has been engulfed by an adjacent cell in vivo. Thus, bona fide 'dead cells' would be different from 'dying cells' that have not yet concluded their demise. In particular, cells that are arrested in the cell cycle (as it occurs during senescence) should be considered as alive, and the expression 'replicative cell death' (which alludes to the loss of clonogenic potential), as it is frequently used by radiobiologists, should be abandoned.

Apoptosis

The expression 'apoptosis' has been coined by Kerr et al. to describe a specific morphological aspect of cell death. Apoptosis is accompanied by rounding-up of the cell, retraction of pseudopodes, reduction of cellular volume (pyknosis), chromatin condensation, nuclear fragmentation (karyorrhexis), classically little or no ultrastructural modifications of cytoplasmic organelles, plasma membrane blebbing (but maintenance of its integrity until the final stages of the process) and engulfment by resident phagocytes (in vivo). Hence, the term 'apoptosis' should be applied exclusively to cell death events that occur while manifesting several among these morphological features. It is worth noting that it is not correct to assume that 'programmed cell death' (PCD) and 'apoptosis' are synonyms because cell death, as it occurs during physiological development, can manifest non-apoptotic features.

Table: Distinct modalities of cell death.

Cell Death Mode	Morphological Features	Notes
Apoptosis	Rounding-up of the cell Retraction of pseudopodes Reduction of cellular and nuclear volume (pyknosis) Nuclear fragmentation (karyorrhexis) Minor modification of cytoplasmic organelles Plasma membrane blebbing Engulfment by resident phagocytes, *in vivo*.	'Apoptosis' is the original term introduced by Kerr *et al.* to define a type of cell death with specific morphological features. Apoptosis is NOT a synonym of programmed cell death or caspase activation.
Autophagy	Lack of chromatin condensation Massive vacuolization of the cytoplasm Accumulation of (double-membraned) autophagic vacuoles Little or no uptake by phagocytic cells, *in vivo*.	'Autophagic cell death' defines cell death occurring with autophagy, though it may misleadingly suggest a form of death occurring by autophagy as this process often promotes cell survival.
Cornification	Elimination of cytosolic organelles Modifications of plasma membrane Accumulation of lipids in F and L granules Extrusion of lipids in the extracellular space Desquamation (loss of corneocytes) by protease activation.	'Cornified envelope' formation or 'keratinization' is specific of the skin to create a barrier function. Although apoptosis can be induced by injury in the basal epidermal layer (e.g., UV irradiation), cornification is exclusive of the upper layers (granular layer and stratum corneum).
Necrosis	Cytoplasmic swelling (oncosis) Rupture of plasma membrane Swelling of cytoplasmic organelles Moderate chromatin condensation.	'Necrosis' identifies, in a negative fashion, cell death lacking the features of apoptosis or autophagy. Note that necrosis can occur in a regulated fashion, involving a precise sequence of signals.

Specific biochemical analyses (such as DNA ladders) should not be employed as an exclusive means to define apoptosis, because this type of cell death can occur without oligonucleosomal DNA fragmentation. Similarly, the presence of proteolytically active caspases or of cleavage products of their substrates is not sufficient to define apoptosis. Frequently, the active suppression (by pharmacological and genetic means) of DNA fragmentation and caspase activation demonstrates that these changes are not required for the execution of the cell death program, although caspase activation may be necessary for the acquisition of the apoptotic morphology. Moreover, the presence of active caspases and of specific products of their enzymatic activity can be linked to non-lethal biological processes. The measurement of DNA fragmentation and of caspase activation, however, may be helpful in diagnosing apoptosis. Thus, it may be reasonable to use caspase activation not only to diagnose but also to better define (together with other features) the type of cell death.

It should be noted that the expression 'apoptosis' hides a major degree of biochemical and functional heterogeneity. There are several distinct subtypes of apoptosis that, although morphologically similar, can be triggered through different biochemical routes (for instance through the 'intrinsic' or the 'extrinsic' pathway, with or without the contribution of mitochondria, etc.). Moreover, the apparent uniformity of apoptotic cell death may conceal heterogeneous functional aspects, for instance concerning the perception of apoptosis by the immune system. Thus, although apoptosis mostly occurs in a non-immunogenic fashion, some lethal stimuli can lead to the exposure or secretion of proteins that elicit the engulfment of apoptotic material by dendritic cells, followed by efficient antigen presentation and stimulation of a specific immune response.

Cell death is frequently considered to be 'caspase-dependent' when it is suppressed by broad-spectrum caspase inhibitors such as N-benzyloxycarbonyl-Val-Ala-Asp-fluoromethylketone (Z-VAD-fmk). As a word of caution, however, it should be noted that Z-VAD-fmk does not act on all caspases with an equal efficiency, and it also inhibits calpains and cathepsins, especially at high concentrations (>10 μM). Moreover, Z-VAD-fmk has been associated with several off-target effects that would result from the binding to cysteines on proteins other than cysteine proteases. As an example, Z-VAD-fmk has been shown to interfere with the interaction between the adenine nucleotide translocase and cyclophylin D, thereby favoring necrotic cell death. For these reasons, the term 'Z-VAD-fmk-inhibitable' should be preferred to 'caspase-dependent'. A second difficulty arises from the fact that caspase inhibition often prevents the appearance of some morphological signs of apoptosis (such as chromatin condensation and DNA fragmentation), yet only retards cell death. In many instances, caspase inhibition simply induces a shift from an apoptotic to a mixed cell death morphology, or even to full-blown pictures of necrosis or autophagic cell death, which, however, may manifest some delay. Thus, 'caspase-independent cell death' can occur despite the efficient inhibition of caspases and can exhibit some of the morphological signs of apoptosis (such as a partial chromatin condensation), autophagy or necrosis.

Considerations on 'Autophagy' and 'Autophagic Cell Death'

Macroautophagy is characterized by the sequestration of cytoplasmic material within autophagosomes for bulk degradation by lysosomes. Autophagosomes, by definition, are two-membraned and contain degenerating cytoplasmic organelles or cytosol, which allows them to be distinguished by transmission electron microscopy from other types of vesicles such as endosomes, lysosomes or apoptotic blebs. The fusion between autophagosomes and lysosomes generates autolysosomes, in which both the autophagosome inner membrane and its luminal content are degraded by acidic lysosomal hydrolases. This catabolic process marks the completion of the autophagic pathway. When the fusion of autophagosomes with lysosomes is blocked, the former accumulate in spite of autophagy inhibition. Hence, a massive increase in the number of autophagosomes is by no means a demonstration that the autophagic pathway is induced, and functional tests are required to investigate autophagy. A very comprehensive description of the assays for monitoring autophagy in higher eukaryotes and a set of guidelines for their interpretation has been recently provided by Klionsky et al. One technique commonly employed to detect autophagy relies on the redistribution of GFP-LC3 fusion proteins into vesicular structures (which can be autophagosomes or autolysosomes). However, the exclusive use of GFP-LC3 as a marker of autophagy is not sufficient to diagnose an enhanced autophagic catabolism.

'Autophagic cell death' is morphologically defined (especially by transmission electron microscopy) as a type of cell death that occurs in the absence of chromatin condensation but accompanied by massive autophagic vacuolization of the cytoplasm. In contrast to apoptotic cells (whose clearance is ensured by engulfment and lysosomal degradation), cells that die with an autophagic morphology have little or no association with phagocytes. Although the expression 'autophagic cell death' is a linguistic invitation to believe that cell death is executed by autophagy, the term simply describes cell death with autophagy. Thus far, involuting Drosophila melanogaster salivary glands provide the only in vivo evidence that the knockdown/knockout of genes required for autophagy truly reduces cell death. This may be due to the limited number of studies that have investigated autophagic cell death in vivo, although there are no doubts that autophagy promotes cell survival, in multiple physiological and experimental settings. Significantly, some reports indicate that cells presenting features of 'autophagic cell death' can still recover upon withdrawal of the death-inducing stimulus. In most cases described to date in which autophagy is suppressed by genetic knockout/knockdown of essential autophagy (*atg*) genes, cell death is not inhibited but rather occurs at an accelerated pace, pointing to the prominent role of autophagy as a pro-survival pathway. This said, it should be noted that most of these studies have been performed on immortalized cell lines *in vitro* and that autophagic cell death rarely affects individual cells *in vivo*. Nevertheless, in specific cases, autophagy may participate in the destruction of cells, as a result of a protracted atrophy of the cytoplasm, beyond a not yet clearly defined point-of-noreturn. Thus, direct induction of autophagy by overexpression of the Atg1 kinase is sufficient to kill fat and salivary gland cells in *Drosophila*. Interestingly, although Atg1-driven autophagic cell death entails caspase-dependent mechanisms in fat cells, the same does not hold true in salivary gland cells (which cannot be rescued from Atg1-induced death by p35 expression).

Necrosis

'Necrotic cell death' or 'necrosis' is morphologically characterized by a gain in cell volume (oncosis), swelling of organelles, plasma membrane rupture and subsequent loss of intracellular contents. For a long time, necrosis has been considered merely as an accidental uncontrolled form of cell death, but evidence is accumulating that the execution of necrotic cell death may be finely regulated by a set of signal transduction pathways and catabolic mechanisms. For instance, death domain receptors (e.g., TNFR1, Fas/CD95 and TRAIL-R) and Toll-like receptors (e.g., TLR3 and TLR4) have been shown to elicit necrosis, in particular in the presence of caspase inhibitors. TNFR1-, Fas/CD95-, TRAILR- and TLR3-mediated cell death seemingly depends on the kinase RIP1, as this has been demonstrated by its knockout/knockdown and chemical inhibition with necrostatin-1. Although there is no generalized consensus on the use of this expression, some authors have proposed the term 'necroptosis' to indicate regulated (as opposed to accidental) necrosis. At a biochemical level, necroptosis may be defined as a type of cell death that can be avoided by inhibiting RIP1 (either through genetic or pharmacological methods), which may represent a convenient means to discriminate between programmed and fortuitous forms of necrosis.

Several mediators, organelles and cellular processes have been implicated in necrotic cell death, but it is still unclear how they interrelate with each other. The causative elements of necrosis are unclear, as well as its bystander effects. These phenomena include mitochondrial alterations (e.g., uncoupling, production of reactive oxygen species, i.e., ROS, nitroxidative stress by nitric oxide or similar compounds and mitochondrial membrane permeabilization, i.e., MMP, often controlled by cyclophilin D), lysosomal changes (ROS production by Fenton reactions, lysosomal membrane permeabilization), nuclear changes (hyperactivation of PARP-1 and concomitant hydrolysis of NAD^+), lipid degradation (following the activation of phospholipases, lipoxygenases and sphingomyelinases), increases in the cytosolic concentration of calcium (Ca^{2+}) that result in mitochondrial overload and activation of non-caspase proteases (e.g., calpains and cathepsins). In several (but not all) instances of necrotic cell death, a crucial role for the serine/threonine kinase RIP1 has been demonstrated. Thus far, however, there is no consensus on the biochemical changes that may be used to unequivocally identify necrosis. In the absence of a common biochemical denominator, necrotic cell death is still largely identified in negative terms by the absence of apoptotic or autophagic markers, in particular when the cells undergo early plasma membrane permeabilization (as compared with its delayed occurrence, which is associated with late-stage apoptosis). For these reasons, caution should be used in classifying particular cell death routines as necrotic.

Cornification

Cornification is a very specific form of PCD that occurs in the epidermis, morphologically and biochemically distinct from apoptosis. It leads to the formation of corneocytes, that is dead keratinocytes containing an amalgam of specific proteins (e.g., keratin, loricrin, SPR and involucrin) and lipids (e.g., fatty acids and ceramides), which are necessary for the function of the cornified skin layer (mechanical resistance, elasticity, water repellence and structural stability). Cornification is less often referred to as 'keratinization' or 'cornified envelope formation', and it is generally considered as a terminal differentiation program similar to those leading to other anucleated tissues (such as the lens epithelium and mature red blood cells). This is mainly due to the fact that these processes display the (often limited) activation of the molecular machinery for cell death, in particular of caspases. In contrast with corneocytes, however, both mature red blood and lens epithelial cells retain the ability to undergo stress-induced death, and hence only cornification should be regarded as a *bona fide* cell death program.

At the molecular level, cornification follows a specific mechanism of epithelial differentiation during which cells express all enzymes and substrates required for building up the epidermal barrier that allows for isolating the body from the external environment. This is obtained by the cross-linking enzymes (e.g., transglutaminase types 1, 3 and 5) acting on several substrates (e.g., loricrin, SPR, involucrin and SP100), as well as through the synthesis of specific lipids that are released into the extracellular space (where they are covalently attached to cornified envelope proteins), and proteases, which are required for impermeability and desquamation, respectively.

Atypical Cell Death Modalities

Mitotic Catastrophe

Mitotic catastrophe is a cell death mode occurring either during or shortly after a dysregulated/failed mitosis and can be accompanied by morphological alterations including micronucleation (which often results from chromosomes and chromosome fragments that have not been distributed evenly between daughter nuclei) and multinucleation (the presence of two or more nuclei with similar or heterogeneous sizes, deriving from a deficient separation during cytokinesis). However, there is no broad consensus on the use of this term, and mitotic catastrophe can lead either to an apoptotic morphology or to necrosis. As a result, the NCDD recommends the use of expressions such as 'cell death preceded by multinucleation' or 'cell death occurring during metaphase', which are more precise and more informative.

Anoikis

Apoptosis induced by the loss of the attachment to the substrate or to other cells is called anoikis. Besides its specific form of induction, the molecular mechanisms of anoikis-associated cell death match those activated during classical apoptosis. The NCCD acknowledges the use of this term for historical reasons, as it is already quite diffuse in the literature. However, it will be necessary to determine whether under certain circumstances other modalities of cell death occur *in vivo* following detachment, that is, whether there are forms of anoikis refractory to caspase inhibitors and others that manifest necrotic features.

Excitotoxicity

This is a form of cell death occurring in neurons challenged with excitatory amino acids, such as glutamate, that leads to the opening of the N-methyl-d-aspartate Ca^{2+}-permeable channel, followed by cytosolic Ca^{2+} overload and activation of lethal signaling pathways. Excitotoxicity seemingly overlaps with other types of death such as apoptosis and necrosis (depending on the intensity of the initiating stimulus), and involves MMP as a critical event. For these reasons, and for the presence of common regulators such as nitric oxide itself, excitotoxicity cannot be considered as a separate cell death modality.

Wallerian Degeneration

Additional less-characterized forms of cellular catabolism take place in the nervous system, such as Wallerian degeneration, in which part of a neuron or axon degenerates without affecting the main cell body. This term does not describe a type of cell death *sensu stricto*, because neurons affected by Wallerian degeneration remain alive.

Paraptosis

This term was originally introduced to describe a form of PCD morphologically and biochemically distinct from apoptosis. In multiple cell types, paraptosis was triggered by the expression of the insulin-like growth factor receptor I, and it was associated with extensive cytoplasmic vacuolization and mitochondrial swelling, but without any other morphological hallmark of apoptosis. The manifestations of paraptosis could not be prevented by caspase inhibitors, nor by the overexpression of antiapoptotic Bcl-2-like proteins, and seemingly resulted from a signaling cascade involving specific members of the mitogen-activated protein kinase family. At present, it is still unclear whether paraptosis represents a route of cell death that is truly distinct from all others.

Pyroptosis

Pyroptosis has first been described in macrophages infected with *Salmonella typhimurium*. It involves the apical activation of caspase-1 (but not of caspase-3), a protease that is mostly known as interleukin-1β(IL-1β)-converting enzyme. Caspase-1 activation induced by *S. typhimurium* (and by other pathogens such as *Pseudomonas aeruginosa* and *Shigella flexneri*) occurs through Ipaf, an Apaf-1-related NLR protein. In contrast, pyroptosis induced by *Bacillus anthracis* lethal toxin does not require Ipaf and rather involves another NLR protein, that is Nalp1. In addition, lipopolysaccharide-treated macrophages (either in the presence or in the absence of ATP) undergo pyroptosis mediated by the adaptor protein ASC, which together with caspase-1 forms a supramolecular cytoplasmic complex also known as 'pyroptosome'. Thus, distinct routes to caspase-1 activation induce pyroptosis. As this form of cell death leads to the release of IL-1β (which is one of the major fever-inducing cytokines or pyrogens) and of IL-18, it may play a relevant role in both local and systemic inflammatory reactions. As it stands, macrophages undergoing pyroptosis not only exhibit morphological features that are typical of apoptosis, but also display some traits associated with necrosis.

Pyronecrosis

Nalp3 and ASC are involved in the necrotic cell death of macrophages infected by *S. flexneri* at high bacteria/macrophage ratios and associated with the release of HMGB-1, caspase-1 and IL-1β, which is called pyronecrosis. Pyronecrosis and pyroptosis are distinguished based on the fact that the latter (but not the former) requires caspase-1. It remains to be determined whether RIP1 is implicated in pyronecrosis, as well as whether pyroptosis and pyronecrosis play any role outside of the innate immune system.

Entosis

Entosis, originally described as a form of 'cellular cannibalism' in lymphoblasts from patients with Huntington's disease, has been reported as a new cell death modality in which one cell engulfs one of its live neighbors, which then dies within the phagosome. Intriguingly, the most efficient cells in performing entosis are MCF-7 breast cancer cells, which lack both caspase-3 and beclin-1 and hence are (relatively) apoptosis- and autophagy-incompetent. This points to the possibility, which remains to be explored, that entosis is a default pathway that is unmasked (and hence can be observed) exclusively when other catabolic reactions are suppressed. Entosis is not inhibited by Bcl-2 or Z-VAD-fmk, and internalized cells appear virtually normal. Later they disappear, presumably through lysosomal degradation. In rare cases, however, internalized cells are able to divide within the engulfing cell or are released. Hence, it is difficult to know whether the cell-in-cell morphology (entosis) truly represents a novel cell death modality.

Postface

As it stands, three distinct routes of cellular catabolism can be defined according to morphological criteria, namely apoptosis (which is a form of cell death), autophagy (which causes the destruction of a part of the cytoplasm, but mostly avoids cell death) and necrosis (which is another form of cell death). Although frequently employed in the past, the use of Roman numerals (i.e., type I, type II and type III cell death, respectively) to indicate these catabolic processes should be abandoned. Moreover, several critiques can be formulated against the clear-cut distinction of different cell types in the triad of apoptosis, autophagic cell death and necrosis.

First, although this vocabulary was originally introduced based on observations of developing animals, it has rapidly been adopted to describe the results of in vitro studies performed on immortalized cell lines, which reflect very poorly the physiology of cell death in vivo. In tissues, indeed, dying cells are usually engulfed well before signs of advanced apoptosis or necrosis become detectable. Thus, it may be acceptable - if the irreversibility of these phenomena is demonstrated - to assess caspase activation and DNA fragmentation to diagnose apoptotic cell death in vivo.

Second, there are numerous examples in which cell death displays mixed features, for instance with signs of both apoptosis and necrosis, a fact that lead to the introduction of terms like 'necroapoptosis' and 'aponecrosis' (whose use is discouraged by the NCCD to avoid further confusion). Similarly, in the involuting D. melanogaster salivary gland, autophagic vacuolization is synchronized with signs of apoptosis, and results from genetic studies indicate that caspases and autophagy act in an additive manner to ensure cell death in this setting. Altogether, these data argue against a clear-cut and absolute distinction between different forms of cell death based on morphological criteria.

Third (and most important), it would be a *desideratum* to replace morphological aspects with biochemical/functional criteria to classify cell death modalities. Unfortunately, there is no clear equivalence between morphology and biochemistry, suggesting that the ancient morphological terms are doomed to disappear and to be replaced by truly biochemical definitions. In this context, 'loss-of-function' and 'gain-of function' genetic approaches (e.g., RNA interference, knockout models and plasmid-driven overexpression systems) represent invaluable tools to characterize cell death modes with more precision, but only if such interventions truly reduce/augment the rate of death, instead of changing its morphological appearance (as it is often the case). Present cell death classifications are reminiscent of the categorization of tumors that has been elaborated by pathologists over the last one and a half centuries. As old morphological categorizations of tumors are being more and more supported (and will presumably be replaced) by molecular diagnostics (which allows for a more sophisticated stratification of cancer subtypes based on molecular criteria), the current catalog of cell death types is destined to lose its value as compared with biochemical/functional tests. In the end, such efforts of classification are only justified when they have a prognostic and predictive impact, allowing the matching of each individual cancer with the appropriate therapy. Similarly, a cell death nomenclature will be considered useful only if it predicts the possibilities to pharmacologically/genetically modulate (induce or inhibit) cell death and if it predicts the consequences of cell death *in vivo*, with regard to inflammation and recognition by the immune system.

Table: Biochemical aspects of distinct modalities of cellular catabolism.

Apoptosis	
Activation of proapoptotic Bcl-2 family proteins (e.g., Bax, Bak, Bid)	If microscopy localization studies. Immunoblotting with conformation-specific antibodies.
Activation of caspases	Colorimetric/fluorogenic substrate-based assays in live cells. Colorimetric/fluorogenic substrate-based assays of lysates in microtiter plates. FACS/IF microscopy quantification with antibodies specifically recognizing the active form of caspases. FACS/IF microscopy quantification with antibodies specific for cleaved caspase substrates. FACS/IF microscopy quantification with fluorogenic substrates. Immunoblotting assessment of caspase-activation state. Immunoblotting assessment of the cleavage of caspase products.
$\Delta\Psi_m$ dissipation	Calcein-cobalt technique (FACS/IF microscopy). FACS/IF microscopy quantification with $\Delta\Psi_m$-sensitive probes. Oxygen-consumption studies (polarography).
MMP	Colorimetric techniques to assess the accessibility of exogenous substrates to IM-embedded enzymatic activities. FACS-assisted detection of IMS proteins upon plasma membrane permeabilization. FACS-assisted detection of physical parameters of purified mitochondria. HPLC-assisted quantification of mitochondrial alterations in purified mitochondria. If microscopy colocalization studies of IMS proteins (e.g., Cyt *c*) with sessile mitochondrial proteins (e.g., VDAC1). IF (video) microscopy with Cyt *c*-GFP fusion protein Immunoblotting detection of IMS proteins (e.g., Cyt *c*) upon cellular fractionation.
Oligonucleosomal DNA fragmentation	DNA ladders FACS quantification of hypodiploid cells (sub-G_1 peak) TUNEL assays

Plasma membrane rupture	Colorimetric/fluorogenic substrate-based assays of culture supernatants in microtiter plates to determine the release of cytosolic enzymatic activities (e.g., LDH). FACS quantification with vital dyes.
PS exposure	FACS quantification of Annexin V binding.
ROS overgeneration	FACS/IF microscopy quantification with ROS-sensitive probes.
ssDNA accumulation	FACS quantification with ssDNA-specific antibodies.

Autophagy	
Beclin-1 dissociation from Bcl-2/X_L	Co-immunoprecipitation studies.
Dependency on *atg* gene products	Genetic studies (e.g., knockout models, RNA interference, plasmid-driven overexpression systems).
LC3-I to LC3-II conversion	IF microscopy with GFP-LC3 fusion protein. Immunoblotting with LC3-specific antibodies.
p62^{Lck} degradation	Immunoblotting with p62-specific antibodies.

Cornification	
Expression of TGs	Immunoblotting with antibodies specific for TG type 1, 3 and 5 qRT-PCR.
Expression of TG substrates	Immunoblotting with antibodies specific for TG substrates (e.g., loricrin, SPR, involucrin, keratins) qRT-PCR.
Crosslinking activity	HPLC detection of K-L isodipeptide bonds. Monodansyl-cadaverine incorporation to detect TG activity in tissues. Radiolabeled putrescine incorporation to detect TG activity in cell extracts.

Necrosis	
Activation of calpains	Colorimetric/fluorogenic substrate-based assays of cell lysates in microtiter plates.
Activation of cathepsins	Colorimetric/fluorogenic substrate-based assays in live cells. Colorimetric/fluorogenic substrate-based assays of cell lysates in microtiter plates.
Drop of ATP levels	Luminometric assessments of ATP/ADP ratio.
HMGB-1 release	Immunoblotting of culture medium with HMGB-1-specific antibodies.
LMP	FACS quantification with lysomorphotropic probes.
Plasma membrane rupture	Colorimetric/fluorogenic substrate-based assays of culture supernatants in microtiter plates to determine the release of cytosolic enzymatic activities (e.g., LDH). FACS quantification with vital dyes.
RIP1 phosphorylation	Immunoblotting with phosphoneoepitope-specific antibodies.
RIP1 ubiquitination	Immunoprecipitation with anti-RIP1 antibodies followed by immunoblotting with anti-ubiquitin antibodies.
ROS overgeneration	FACS quantification with ROS-sensitive probes.
Specific PARP1 cleavage pattern	Immunoblotting with PARP1-specific antibodies.

Abbreviations: $\Delta\Psi_m$, mitochondrial transmembrane permeabilization; Cyt c, cytochrome c; FACS, fluorescence-activated cell sorter; GFP, green fluorescent protein; HPLC, high-pressure liquid chromatography: IF, immunofluorescence; IM, mitochondrial inner membrane; IMS, mitochondrial intermembrane space; LDH, lactate dehydrogenase; LMP, lysosomal membrane permeabilization; MMP, mitochondrial membrane permeabilization; PS, phosphatidylserine; qRT-PCR, real-time quantitative reverse transcription PCR; ROS, reactive oxygen species; RNAi, RNA interference; TG, transglutaminase; TUNEL, terminal deoxynucleotidyl transferase-mediated dUTP nick-end labeling; VDAC1, voltage-dependent anion channel 1.

References

- Cellularrespiration: biology-pages.info, Retrieved 15 January, 2019

- Aerobic-respiration: biologydictionary.net, Retrieved 21 April, 2019

- Cell-reproduction: biologyfunfacts.weebly.com, Retrieved 15 June, 2019

- Cell-division: biologydictionary.net, Retrieved 18 May, 2019

- Binary-fission: microscopemaster.com, Retrieved, 21 July, 2019

- Biobookmeiosis: estrellamountain.edu, Retrieved, 17 February, 2019

- The-cell-cycle, biology: lumenlearning.com, Retrieved, 24 March, 2019

- Cell-transport-and-homeostasis: opencurriculum.org, Retrieved 1 August,2019

- Cytoplasmic-streaming, science: britannica.com, Retrieved 21 July, 2019

- Cellular-senescence-what-why-and-how: woundsresearch.com, Retrieved 15 May, 2019

Permissions

Index

www.ingramcontent.com/pod-product-compliance
Lightning Source LLC
Chambersburg PA
CBHW082050190326
41458CB00010B/3501